BeagleBone Robotic Projects

Create complex and exciting robotic projects with
the BeagleBone Black

Richard Grimmett

BIRMINGHAM - MUMBAI

BeagleBone Robotic Projects

First published: December 2013

Production Reference: 1181213

Published by Packt Publishing Ltd.

Livery Place

35 Livery Street

Birmingham B3 2PB, UK.

ISBN 978-1-78355-932-9

www.packtpub.com

Cover Image by Disha Haria (dishah@packtpub.com)

Credits

Author
Richard Grimmett

Reviewers
Álvaro García Gómez

Lihang Li

Derek Molloy

Acquisition Editor
Sam Birch

Lead Technical Editor
Chalini Snega Victor

Technical Editors
Jalasha D'costa

Monica John

Edwin Moses

Nikhil Potdukhe

Siddhi Rane

Sonali S. Vernekar

Project Coordinator
Leena Purkait

Proofreader
Chris Smith

Indexer
Tejal Soni

Graphics
Sheetal Aute

Abhinash Sahu

Production Coordinators
Alwin Roy

Kirtee Shingan

Cover Work
Kirtee Shingan

About the Author

Richard Grimmett has always been fascinated by computers and electronics from his very first programming project that used Fortran on punch cards. He has a Bachelor's and Master's degree in Electrical Engineering and a PhD in Leadership Studies. He also has 26 years of experience in the Radar and Telecommunications industries, and even has one of the original brick phones. He now teaches Computer Science and Electrical Engineering at Brigham Young University - Idaho where his office is filled with many of his robotics projects.

I would certainly like to thank my wife and family for providing me the time and wonderful, supportive environment that encourages me to take on projects such as this one. I would also like to thank my students; they always amaze and inspire me with their creativity when released from the boredom of standard educational practices.

About the Reviewers

Álvaro García Gómez is a computer engineer at the University of Valladolid (Spain) and a technical administrator of IT systems. He was focused on software development, but a short time later robotics and embedded devices aroused his curiosity. Now he is specialized in machine learning and autonomous robotics, which involve his two passions: computing and electronics. Now he is working in his own company that develops free software and hardware.

Lihang Li received his B.E. degree in Mechanical Engineering from Huazhong University of Science and Technology (HUST), China in 2012 and is now pursuing his M.S. degree in Computer Vision at National Laboratory of Pattern Recognition (NLPR) from the Institute of Automation, Chinese Academy of Sciences (IACAS).

He is a member of Dian Group from HUST and mainly concentrated on Embedded System Development when he was an undergraduate. He is familiar with Embedded Linux, ARM, DSP, and various communication interfaces (I2C, SPI, UART, CAN, and ZigBee, among others). He took part in a competition called The Asia-Pacific Robot Contest (ABU Robocon) with his team in 2012 and secured third place among 29 teams in China.

As a graduate student, he is focusing on Computer Vision and specially on SLAM algorithms. In his free time, he likes to take part in Open Source Activities and now is President of the Open Source Club, Chinese Academy of Sciences. Also, building a multicopter is his hobby and he is with a team called OpenDrone from Beijing Linux User Group (BLUG).

His interest includes: Linux, Open Source, Cloud Computing, Virtualization, Computer Vision algorithms, Machine Learning and Data Mining, and various programming languages.

You can find him at his personal website, http://hustcalm.me.

Many thanks to my girlfriend Jingjing Shao; it was her encouragement to push me to be a reviewer for this book. And I appreciate her kindness though sometimes I can't spare time for her. Also, I must thank all the team: Leena, who is a very good Project Coordinator, and the other reviewers, though we haven't met, I'm happy to work with you.

Derek Molloy is a senior lecturer in the School of Electronic Engineering, Faculty of Engineering & Computing at Dublin City University, Ireland. Since 1997, he has lectured in object-oriented programming, 3D Computer Graphics, and Digital Electronics at postgraduate and undergraduate levels. His research interests are in the fields of Computer & Machine Vision, 3D Graphics and Visualization, and e-learning. He is a key academic member of the Centre for Image Processing and Analysis (CIPA) at DCU. He has published his works widely in international journals and conferences, including an important textbook, *Machine Vision Algorithms in Java, Springer* (2001). In his spare time he runs the DerekMolloyDCU YouTube channel that contains many instructional videos on the use of the BeagleBone, and he integrates everything on his personal blog at www.derekmolloy.ie.

www.PacktPub.com

Support files, eBooks, discount offers, and more

You might want to visit `www.PacktPub.com` for support files and downloads related to your book.

Did you know that Packt offers eBook versions of every book published, with PDF and ePub files available? You can upgrade to the eBook version at `www.PacktPub.com` and as a print book customer, you are entitled to a discount on the eBook copy. Get in touch with us at `service@packtpub.com` for more details.

At `www.PacktPub.com`, you can also read a collection of free technical articles, sign up for a range of free newsletters and receive exclusive discounts and offers on Packt books and eBooks.

`http://PacktLib.PacktPub.com`

Do you need instant solutions to your IT questions? PacktLib is Packt's online digital book library. Here, you can access, read and search across Packt's entire library of books.

Why Subscribe?

- Fully searchable across every book published by Packt
- Copy and paste, print and bookmark content
- On demand and accessible via web browser

Free Access for Packt account holders

If you have an account with Packt at `www.PacktPub.com`, you can use this to access PacktLib today and view nine entirely free books. Simply use your login credentials for immediate access.

Table of Contents

Preface	**1**
Chapter 1: Getting Started with the BeagleBone Black	**9**
Mission briefing	9
The unveiling!	11
Hooking up a keyboard, mouse, and display	15
Changing the operating system	18
Adding a graphical user interface	22
Accessing the board remotely	26
Mission accomplished	34
A challenge	34
Chapter 2: Programming the BeagleBone Black	**35**
Mission briefing	35
Basic Linux commands and navigating the filesystem	36
Creating, editing, and saving files on the BeagleBone Black	42
Creating and running Python programs on the BeagleBone Black	43
Basic programming constructs on the BeagleBone Black	47
Introduction to the C++ programming language	54
Mission accomplished	59
A challenge	59
Chapter 3: Providing Speech Input and Output	**61**
Mission briefing	61
Hooking up the HW to make and input sound	64
Using eSpeak to allow your projects to respond in a robotic voice	70
Using PocketSphinx to interpret your commands	73
Providing the capability to interpret your commands and have your robot initiate an action	80

Mission accomplished   83
A challenge   83

Chapter 4: Allowing the BeagleBone Black to See   85
Mission briefing   85
Connecting the USB camera to the BeagleBone Black and viewing the images   86
Downloading and installing OpenCV – a full-featured vision library   89
Using the vision library to detect colored objects   97
Mission accomplished   102
Challenges   102

Chapter 5: Making the Unit Mobile – Controlling Wheeled Movement   103
Mission briefing   103
Using a motor controller to control the speed of your platform   107
Controlling your mobile platform programmatically using the BeagleBone Black   117
Making your mobile platform truly mobile by issuing voice commands   122
Mission accomplished   124
A challenge   124

Chapter 6: Making the Unit Very Mobile – Controlling Legged Movement   125
Mission briefing   125
Connecting the BeagleBone Black to the mobile platform using a servo controller   130
Creating a program in Linux to control the mobile platform   138
Making your mobile platform truly mobile by issuing voice commands   142
Mission accomplished   143
A challenge   144

Chapter 7: Avoiding Obstacles Using Sensors   145
Mission briefing   145
Connecting the BeagleBone Black to a USB sonar sensor   148
Using a servo to move a single sensor   154
Mission accomplished   159
A challenge   159

Chapter 8: Going Truly Mobile – Remote Control of Your Robot   161
Mission briefing   161
Connecting the BeagleBone Black to a wireless USB keyboard   167
Using the keyboard to control your project   169
Mission accomplished   174
A challenge   174

Chapter 9: Using a GPS Receiver to Locate Your Robot   175
Mission briefing   175
Connecting the BeagleBone Black to a GPS device   176
Accessing the GPS programmatically and determining

how to move to a location   188
Mission accomplished   193
A challenge   193

Chapter 10: System Dynamics   195
Mission briefing   195
Creating a general control structure so capabilities can communicate 197
Mission accomplished   205
A challenge   206

Chapter 11: By Land, Sea, and Air   207
Mission briefing   207
Using the BeagleBone Black in sailing robots   208
Using the BeagleBone Black in flying robots   216
Using the BeagleBone Black in submarine robots   222
Mission accomplished   224
A challenge   224

Index   225

Preface

We live in an amazing age. We are mostly aware of how amazing it is as we live in an age where major changes to how we live occur well within a lifetime, sometimes within a few years. Nowhere is this more evident than in the general area of technology, and the specific area of computers. Not so many years ago, certainly within the lifespan of most of the baby-boomer generation, computers were distant machines kept in the backrooms of large corporations or universities. Access to them was tightly controlled. If you wanted to program them, you punched your computer cards, fed them into the card reader, and then, after an hour or so of wait, you went to receive your computer printout. This was, I regret to reveal, part of my first experience with a computer.

These large computers were the domain of companies such as IBM, with their model 360, Digital Equipment, with the Model PDP-7, and Hewlett-Packard, with the Model 1000. These computers cost many thousands of dollars, and were rarely seen except by a privileged few, who had access to climate-controlled computer rooms.

This model fit the world just wonderfully for many years, until the advent of the personal computer. I was lucky to know someone who purchased one of the very first IBM-PCs. It had two floppies, a monochrome monitor, and was an amazing piece of equipment. Suddenly the world changed and the technology that had seemed so remote was now available on the desktop. This same technological revolution in processing power also birthed a new breed of dedicated microprocessors. These could be used for specific tasks that had previously been the realm of analog circuitry or, in many cases, human interaction with mechanical systems.

These processing solutions to specific applications are named embedded systems. They take the digital calculating capability of personal computers and shrink them even further so they can be placed in common household and industrial objects. Embedded technology has also evolved with respect to price; fortunate, for few would be willing to pay several thousand dollars for a door lock or temperature sensor. The initial embedded devices were very limited in their technology, and developing applications with them became quite a challenge. It was very common to run out of either computing horsepower or memory. Many nights were spent by the talented few shoehorning the last features into the last few bytes of memory.

The computer age has spawned an amazing array of technical advances in both the hardware and software areas. Companies such as Intel and AMD have created processors with almost unfathomable computing power and more available memory that once thought possible, and both Microsoft and Apple have provided major advances in the area of software functionality and usability. The personal computer has become a standard tool in most households, schools, businesses, and factories.

As the personal computer has gone, so has the embedded systems world. From what were once four bit, special purpose processors with 2000 bytes of memory, now embedded processors have emerged that rival the performance and capability of standard personal computers. One has to look no further than the cell phone for an example of significant computing capability in very small packages, and at very inexpensive prices.

This has all reached a bit of a crescendo with the introduction of small, inexpensive systems that can not only run simple, focused applications, but have the capability of powering almost any type of computing need we can create. At the same time these small but powerful systems have outgrown the small, single purpose development environments as well. They are now paired with powerful operating systems, and provide personal computer-like functionality in very small packages. The overwhelming advance of tablets and smart phones has begun to take over the face of computing for many applications.

These advances have also affected the embedded area as well. Small, highly capable systems have married very inexpensive hardware with free, open source software to provide a platform for almost anyone to explore the embedded world. The Arduino, the Raspberry Pi, and now the BeagleBone Black are all platforms that offer not only an affordable price point, but also an open source software community that provides free capability and an easy way to interact with others to get answers to questions or exchange ideas. With these new capabilities, as we shall see later in the book, the sky is literally the limit.

This book will focus on just one of these processors, the BeagleBone Black. However, much of what is written here could be applied to other choices with some limited modifications. But this is not what you came here to learn. You came to learn how to build some very interesting, complex, amazing robotics projects. And processors such as the BeagleBone Black are impressive because they have the capability to not only make this possible, but to make it accessible to those outside of academic or research communities. In this book, we'll explore these capabilities, and build some very impressive projects.

Just a few comments on how the book is laid out. We'll start with a very basic introduction to the BeagleBone Black, and how to get the hardware and software up and working. Then, we'll build some basic functionality on top of the basic system, showing you how to add sound, vision, and control.

Then we'll tackle some fairly complex capability, including GPS, audio, and some advanced sensors. Finally, we'll wrap it up by showing you how to put an entire system together with some tools that can make that a bit less complicated.

In each chapter, I'll give you some very specific instructions for how to proceed. This is a bit dangerous, and the instructions are all going to be subject to change. Hopefully you'll understand the basics of what we are trying to accomplish, so if things don't go quite to plan, you'll be able to figure out how to proceed. There is a lot of help out there, between message boards and blogs, so don't be shy.

What is critical to remember is that this is not an academic exercise. Don't just read the book, but do something with the hardware. My hope is that by the end, you'll be building the kinds of machines that will lead us all into the 22nd century. I often tell my students that their children will grow up as comfortable with robots as they are with computers.

So, let's begin!

What this book covers

Chapter 1, *Getting Started with the BeagleBone Black*, will provide instructions for initial power-up of your hardware.

Chapter 2, *Programming the BeagleBone Black*, will give you a brief tutorial so that you can be successful implementing all the amazing functionality, as many of you are new to embedded systems, Linux, Python, or perhaps even programming in general.

Chapter 3, *Providing Speech Input and Output*, will show you how to add speech recognition as well as make your robot speak.

Chapter 4, *Allowing the BeagleBone Black to See*, will show you how to add the capability for your robot to see.

Chapter 5, *Making the Unit Mobile – Controlling Wheeled Movement*, will show you how to add wheeled movement to your robot.

Chapter 6, *Making the Unit Very Mobile – Controlling Legged Movement*, shows how to build robots that have the capability to walk.

Chapter 7, *Avoiding Obstacles Using Sensors*, shows how to use sensors to avoid barriers as it hardly makes sense to have mobility if your robot is going to run into obstacles.

Chapter 8, *Going Truly Mobile – Remote Control of Your Robot*, will show how to use a remote device to control your robot.

Chapter 9, *Using a GPS Receiver to Locate Your Robot*, shows how to add a GPS device to your robot.

Chapter 10, *System Dynamics*, introduces some methods for organizing all of the capabilities so that they are available at the same time.

Chapter 11, By Land, Sea, and Air, introduces some interesting possibilities for embedded projects that can fly, sail, or swim

What you need for this book

Each chapter will lead you through not only the hardware, but also the software required for each project. However, for almost all of these projects you'll need a personal computer connected to the Internet, an additional Internet connection and LAN cable, the BeagleBone Black, and the power cable that comes with it.

Who this book is for

This book is designed for the informed beginner. I would hope that before beginning the projects in this book you would be familiar with your personal computer and its basic use and functionality. You won't need prior programming experience, but it will be helpful. You'll be introduced to some of the most basic working of the Linux operating system, so any familiarity there will be helpful, but not essential. More than anything the book requires a curiosity about how robots or other embedded projects work, and the tenacity to work through the issues associated with building your own hardware and then adding software to get to a working system.

Conventions

In this book, you will find several headings appearing frequently.

To give clear instructions of how to complete a procedure or task, we use:

Mission briefing

This section explains what you will build, with a screenshot of the completed project.

Why is it awesome?

This section explains why the project is cool, unique, exciting, and interesting. It describes what advantage the project will give you.

Your objectives

This section explains the major tasks required to complete your project.

- ▸ Task 1
- ▸ Task 2
- ▸ Task 3
- ▸ Task 4, and so on

Mission checklist

This section explains any pre-requisites for the project, such as resources or libraries that need to be downloaded, and so on.

Task 1

This section explains the task that you will perform.

Prepare for lift off

This section explains any preliminary work that you may need to do before beginning work on the task.

Engage thrusters

This section lists the steps required in order to complete the task.

Objective complete – mini debriefing

This section explains how the steps performed in the previous section allow us to complete the task. This section is mandatory.

Classified intel

The extra information in this section is relevant to the task.

You will also find a number of styles of text that distinguish between different kinds of information. Here are some examples of these styles, and an explanation of their meaning.

Code words in text are shown as follows: "You can do this with the `ls -la /dev/sd*` command."

A block of code is set as follows:

```
#Smooth image, then convert the Hue
    cv.Smooth(img,img,cv.CV_BLUR,3)
    hue_img = cv.CreateImage(cv.GetSize(img), 8, 3)
    cv.CvtColor(img,hue_img, cv.CV_BGR2HSV)
```

Any command-line input or output is written as follows:

```
xz -cd ubuntu-precise-12.04.2-armhf-3.8.13-bone20.img.xz > /dev/sdX
```

New terms and **important words** are shown in bold. Words that you see on the screen, in menus or dialog boxes for example, appear in the text like this: "The **Safe start violation** tab is set when you first enter the program; you need to clear this by clicking on the **Resume** button at the bottom-left corner of the screen."

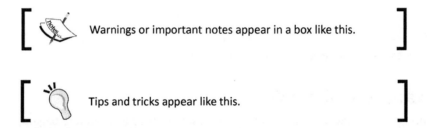

Warnings or important notes appear in a box like this.

Tips and tricks appear like this.

Reader feedback

Feedback from our readers is always welcome. Let us know what you think about this book—what you liked or may have disliked. Reader feedback is important for us to develop titles that you really get the most out of.

To send us general feedback, simply send an e-mail to feedback@packtpub.com, and mention the book title via the subject of your message.

If there is a topic that you have expertise in and you are interested in either writing or contributing to a book, see our author guide on www.packtpub.com/authors.

Customer support

Now that you are the proud owner of a Packt book, we have a number of things to help you to get the most from your purchase.

Downloading the example code and colored images

You can download the example code files and colored images for this Packt book you have purchased from your account at `http://www.packtpub.com`. If you purchased this book elsewhere, you can visit `http://www.packtpub.com/support` and register to have the files e-mailed directly to you.

Errata

Although we have taken every care to ensure the accuracy of our content, mistakes do happen. If you find a mistake in one of our books—maybe a mistake in the text or the code—we would be grateful if you would report this to us. By doing so, you can save other readers from frustration and help us improve subsequent versions of this book. If you find any errata, please report them by visiting `http://www.packtpub.com/submit-errata`, selecting your book, clicking on the **errata submission form** link, and entering the details of your errata. Once your errata are verified, your submission will be accepted and the errata will be uploaded on our website, or added to any list of existing errata, under the Errata section of that title. Any existing errata can be viewed by selecting your title from `http://www.packtpub.com/support`.

Piracy

Piracy of copyright material on the Internet is an ongoing problem across all media. At Packt, we take the protection of our copyright and licenses very seriously. If you come across any illegal copies of our works, in any form, on the Internet, please provide us with the location address or website name immediately so that we can pursue a remedy.

Please contact us at `copyright@packtpub.com` with a link to the suspected pirated material.

We appreciate your help in protecting our authors, and our ability to bring you valuable content.

Questions

You can contact us at `questions@packtpub.com` if you are having a problem with any aspect of the book, and we will do our best to address it.

Getting Started with the BeagleBone Black

Ordering the hardware (HW) is the exciting part of any project. You have wonderful dreams of all that you might accomplish once this amazing piece of technology is delivered. Unfortunately, the frustration of the first few attempts at accessing the capabilities of the unit can leave many developers, especially those with little experience with this type of dedicated system, so discouraged that the board can end up on the shelf, gathering dust next to the pet rock and cassette tape recorder.

Mission briefing

There is rarely anything as exciting as ordering the latest new technology and anticipating its arrival. You daydream of the projects you'll build, the amazing things you can do, the accolades you'll receive from family, friends, and colleagues. However, reality rarely fulfills your fantasies. This project will hopefully help you avoid the pitfalls that normally accompany unboxing and configuring your BeagleBone Black. You'll step through the process, answer all kinds of clarifying questions, and help you understand what is going on. If you don't get through this project, then you'll not be successful at any of the others, so buckle up and get ready for an exciting ride.

The most challenging aspect of accomplishing this for me as your guide is trying to decide to what level I should describe each step. Some of you are beginners, some have limited experience, others will know significantly more than I in some of these areas. I'll try to keep it brief, but also try to be thorough, so that at least you'll know what steps to take in order to be successful. I'll also try to point out some of the different ways you can get help if you are encountering problems. So for this project, here are your objectives.

Your objectives

Your objectives are as follows:

- Hooking up a keyboard, mouse, and display
- Changing the operating system
- Adding a graphical user interface
- Accessing the board remotely

Downloading the example code and colored images

You can download the example code and colored images for this book you have purchased from your account at `http://www.packtpub.com`. If you purchased this book elsewhere, you can visit `http://www.packtpub.com/support` and register to have the files e-mailed directly to you.

Mission checklist

Here are the items you'll need for this project:

- A BeagleBoard Black
- The USB cable provided with the board
- A display with the proper video input
- A keyboard, mouse, and powered USB hub
- A micro SD card of at least 4 GB
- A micro SD card reader/writer that fits your computer
- Another computer that is connected to the Internet
- An Internet connection for the board

The unveiling!

The board has finally arrived. Here is what should come with the standard package:

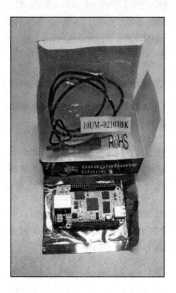

Prepare for lift off

Before plugging anything in, inspect the board for any issues that might have occurred during shipping. This is normally not a problem, but it is always good to do a quick visual inspection. You should also acquaint yourself with the different connections on the board. Here they are, labeled for your information:

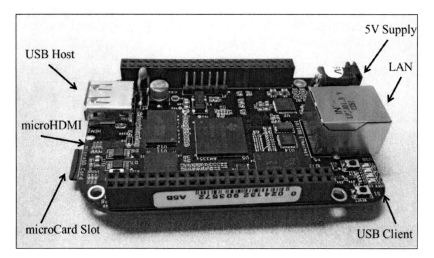

Engage thrusters

So let's get started. You need to power the board, but you also need to hook up a way to interact with the board and see the results of your interaction. The first thing you'll notice is that there is no cable that fits the 5V DC connector. What's with that? Am I already hung up without ever powering on the board? Well, fortunately no, but you do need to talk about power for a moment. There are two ways to power the board. The first is through the USB client connection. This is done by:

 ▶ Connecting the micro-USB connector end of the cable to the board
 ▶ Connecting the standard sized USB connector to either a PC or a compatible DC power source that has such a connection

If you are going to use a DC power source, make sure the unit can supply at least one ampere. This is not optional. Although the board might not always draw this much current, if it senses that the unit cannot supply the required current, it will shut down.

There is another option to power the board: simply supply 5V DC to the connector. Make sure that the plug is 5.5 x 2.1 mm (centre positive) and that the unit can supply at least one ampere. As mentioned earlier, this is not optional.

Even if you are going to choose a DC power source for your board, initially let's connect the board via the provided USB cable. Almost all of the different projects you work on here will need to supply power from a battery pack anyway, and if you supply the board through the USB port and micro-USB connector, you can use your external computer to communicate with the board and ensure that it is up and working.

Objective complete – mini debriefing

When you plug the board in, the **PWR** LED, located by the 5V input, should light blue on the board. Here is a close up of the LED locations, just so that you're certain which one to look for:

The other four indicators on the right-hand side of the LAN connection will eventually begin to flash blue. The one on the far right will eventually flash as a heartbeat indicator, letting you know that your processor is working by flashing twice quickly, approximately once per second.

Now you can use some computer software (SW) to make sure your board is operating correctly. When you first plug the board into a Windows PC, you'll see the indicator at the lower-right indicating that new HW is being installed. Eventually—and this may take a bit—you'll get the indication that your device is ready to use. If you are using Windows 7, you can view the device in your **Devices and Printers** display (select this from the **Start** menu). You should see this:

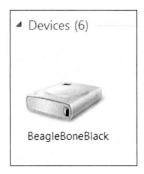

If you see this and the farthest right LED flashing in a heartbeat fashion, you've successfully connected your board. If you can't reach this point, see the following *Classified intel* section.

Once you've connected, you can actually communicate with your board via the USB connection. Open a Firefox or Chrome browser and type in the address `192.168.7.2`. You should see the following in your browser:

If you've reached this point, congratulations! You are now communicating with your BeagleBone Black as the web pages are being served by the on-board web server. You're ready for the next step. Don't continue with directions on this page; you're going to take a different route in updating your BeagleBone Black. If you're having problems, the folks at beagleboard.org have a rich set of forums that can help you work through any of the problems you might be having unpacking the board.

Classified intel

Powering the board can be a bit challenging, since the board requires at least 500 mA of current, and many USB cables and ports are limited by design to less than 500 mA. When attempting to power up with these cables on a power supply that cannot supply enough current, the unit will begin to power on, the blue LEDs will begin to flash, and then everything will turn off. This was a more significant problem with early units than the units that are currently shipping.

Also, if you are struggling to connect to the board, you may need to download drivers. These are available on the beagleboard.org site.

Hooking up a keyboard, mouse, and display

The board is now powered on, and you have blinking LEDs. You have been able to access the basic functionality via the USB port. However, you want to do much more. This task will help you accomplish your goals.

Prepare for lift off

Now that you know that your board works and you've verified the connection via the provided USB cable to your computer, you're going to add the peripherals so that it can operate as a standalone computer system. This step is a bit optional, as in the future your projects will be in systems where you won't connect directly to the board with a keyboard, mouse, and display. However, this can be a great learning step and in the off chance, you need to do some debugging on the system. It is good to know how to connect directly to the board. You'll also use this configuration to verify the basic SW installation before you start your projects.

You'll need:

- A USB mouse.
- A keyboard.
- A display device.
- You may also need a USB hub, and if you don't have one, get a powered USB hub. This will be important later in your project work.

Many of you will have most of this stuff already, but if you don't, there are some things to consider before buying additional equipment. Let's start with the keyboard and mouse. Most mice and keyboards have separate USB connectors. You'll notice, however, that on your BeagleBone Black there is only one USB port; thus, there is a need for the USB hub.

Before deciding on the hub to connect to your board, you need to understand the difference between a powered USB hub, and one that gets its power from the USB port itself. Almost all USB hubs are not powered; that is, you don't plug in the USB hub separately. The reason for this is, almost all of these hubs are hooked up to computers with very large power supplies, and so powering USB devices from the computer is not a problem. This is not the case for your board. The USB port on your board has very limited power capabilities, so if you are going to hook up devices that require significant power—for instance, a WLAN adapter or Kinect sensor—you're going to need a powered USB hub, one that provides power to the devices through a separate power source.

If you already have a hub that is not powered and want to use it to connect your keyboard and mouse, feel free, it should work fine as these do not draw too much power. If you don't already have a keyboard and mouse, or are looking for a keyboard and mouse that you can dedicate to your board, I suggest choosing a keyboard with a mousepad. That way you only have one USB connection to the two devices.

To complete this step you'll also need a display. You need to investigate which display types can be used with the BeagleBone Black. The only video output on the board is a micro-HDMI connector. The easiest connection to create is to connect the board directly to a monitor or TV that has an HDMI input; however, you'll probably need to buy either a cable that has micro-HDMI on one end and regular HDMI on the other, or an adapter from micro-HDMI to regular HDMI. HDMI monitors are relatively new, but if you have a monitor that has a DVI input, you can buy adapters or HDMI to DVI cables relatively inexpensively that provide an interface between DVI and HDMI. The display I use has a DVI input.

Don't be fooled, however, by adapters that claim that they go from HDMI to VGA, or HDMI to s-video. These are two different kinds of signals: HDMI and DVI are digital standards, and VGA and s-video are analog standards. There are adapters that can do this, but they must contain circuitry and require power, so are significantly more expensive than any simple adapter, and they result in a lower quality output.

Engage thrusters

Now that you have your parts, connect the USB hub to the standard USB port, the keyboard and mouse to the USB hub, and the display to the micro-HDMI connector as shown here:

Once these are all connected, plug in the USB hub, the display, and finally the BeagleBone Black board. Since I am no longer going to use the USB connection to the computer, I am using a standard USB 5 volt power supply. Make sure you connect all your devices before you power on the unit. Most operating systems support hot-swap of devices, which means you are able to connect a device after the system has been powered, but this is a bit shaky in the embedded environment. You'll always cycle power when you connect a new HW.

Objective complete – mini debriefing

Once this is complete, the unit should power on; it will boot its default operating system from the internal eMMC space, which is a sort of internal memory card and the display will look like this:

So you should now be able to interact with your BeagleBone Black directly. This is an important step, although for most of your projects you'll use a remote computer to program and control your device. Keep the components you have gathered around for debug purposes, you may need them later, as there is at least a possibility that things might go wrong and you'll need to find out how to fix them.

Classified intel

Just a couple of notes about this task. First, if you have problems powering the system, check to make sure your power supply can supply enough current. Don't even consider powering the system with less than 1 amp. Also, if you are using a power supply with a USB connector, make sure you use the cable that came with the unit to connect between the USB and the unit. Some cables also limit the current, and the unit will sense this and not power on correctly.

Also, a note on connecting to the display: the board's HDMI connector is micro-HDMI, which almost begs for an adapter. The display I chose to use is an inexpensive monitor with DVI input, so I purchased a cable that went from standard HDMI to DVI. Then I purchased a micro-HDMI to standard HDMI adapter. For some reason, I had problems with this configuration, and chalked it up to a bad HDMI adapter. I now prefer a cable that has a micro-HDMI connector on one end, and a standard HDMI connector on the other, and then a standard HDMI to DVI adapter. This seems to be the most reliable and, if I choose, I can quickly change and use my HDTV as the display. One of the challenges in choosing the components for your system is trying to anticipate how it might be used in the future.

Changing the operating system

Now that you have your system all up and working, you're going to do something that seems a bit counterintuitive; instead of using the system you've got, you're going to install a new operating system on card so your board will boot and run this different operating system. The reason will become clear in the next section.

Prepare for lift off

The default operating system on the internal memory is a version of Linux called **Ångström**. Now Linux, unlike Windows, Android or iOS, is not tightly controlled by a single company. It is a group effort, mostly open source, and while it is available for free, it grows and develops a bit more chaotically.

Thus a number of "distributions" have emerged, each built on a similar kernel, or core set of capabilities. These core capabilities are all based on the Linux specification. However, they are packaged slightly differently, and developed, supported, and packaged by different organizations. Ångström is one of these versions; Ubuntu is another. There are others as well, but these are the two main choices for the distribution that you will put on your card.

I choose to use the Ubuntu distribution of Linux on my BeagleBone Black for a couple of reasons. First, Ubuntu is arguably the most popular distribution of Linux, which makes it a good choice because of the community support it offers. Also, I personally like this distribution of Linux when I need to run Linux on my own personal computer. It provides a complete set of features, is well organized, and generally supports the latest sets of HW and SW. Having the same version on both my personal computer and my BeagleBone Black makes it easier for me to use both as they operate, at least to a certain degree, the same way. I can also try some things on my computer before trying them on the BeagleBone Black. I've also found that Ubuntu has excellent support for new HW, and this can be very important for your projects.

Others tend to favor the Ångström distribution, the support for this distribution is growing and it can sometimes be a bit simpler to access and work with. There are also other choices, such as Arch and there are some who are working on a distribution of Android for the BeagleBone Black. But for this book we are going to install and run a version of Ubuntu on your BeagleBone Black.

Engage thrusters

There are two approaches to getting Ubuntu onto your board. The board is getting popular enough that you can buy an SD card that already has Ubuntu installed, or you can download it onto your personal computer and then install it on the card. I'll assume you don't need any directions if you want to purchase a card—simply do an Internet search for companies selling such a product.

If you are going to download a distribution, you need to decide if you are going to use a Windows computer to download and create an SD card, or a Linux machine. I'll give brief directions for both here.

First, you'll need to download an image. This part of the process is similar for either Windows or Linux. Open a browser window. You can go to one of the several sites that have an image you can put on your card. My personal favorite is `http://elinux.org/ Beagleboard:Ubuntu_On_BeagleBone_Black`. They keep pointers to a number of different images and directions on how to install them. My personal favorite is the 12.04 version of Ubuntu. It is new enough to support everything you need, but old enough to be stable. Select the image and download the file.

If you're using Windows, you'll need to unzip the file using an archiving program like 7-Zip. If you don't have this on your computer, follow the directions on the beaglebone.org Getting Started web page. This will leave you with a file that has the `.img` extension, a file that can be imaged on your card.

Now that you have the image, you need a program that can write the image to the card. I use the Image Writer for Windows program. Again, if you don't have this program, follow the directions on the beaglebone.org Getting Started web page. Plug your card into the PC, run this program, select the correct card and image, then select **Write**. This will take some time, but when completed eject the card from the PC.

If you are using Linux, you'll need to un-archive the file and then write it to the card. You can do this all with one command. However, you do need to find the /dev device label for your card. You can do this with the ls -la /dev/sd* command. If you run this before you plug in your card, you might see something like this:

```
richard@vicki-automated: ~
richard@vicki-automated:~$ ls -la /dev/sd*
brw-rw---- 1 root disk 8, 0 Jul  4 10:34 /dev/sda
brw-rw---- 1 root disk 8, 1 Jul  4 10:34 /dev/sda1
brw-rw---- 1 root disk 8, 2 Jul  4 10:34 /dev/sda2
brw-rw---- 1 root disk 8, 5 Jul  4 10:34 /dev/sda5
richard@vicki-automated:~$
```

After plugging in your card, you might see something like this:

```
richard@vicki-automated: ~
richard@vicki-automated:~$ ls -la /dev/sd*
brw-rw---- 1 root disk 8,  0 Jul  4 10:34 /dev/sda
brw-rw---- 1 root disk 8,  1 Jul  4 10:34 /dev/sda1
brw-rw---- 1 root disk 8,  2 Jul  4 10:34 /dev/sda2
brw-rw---- 1 root disk 8,  5 Jul  4 10:34 /dev/sda5
brw-rw---- 1 root disk 8, 16 Jul 11 09:50 /dev/sdb
brw-rw---- 1 root disk 8, 17 Jul 11 09:50 /dev/sdb1
brw-rw---- 1 root disk 8, 18 Jul 11 09:50 /dev/sdb2
richard@vicki-automated:~$
```

Notice your card is at sdb. Now go to the directory where you downloaded the archived image file and issue the following command:

```
xz -cd ubuntu-precise-12.04.2-armhf-3.8.13-bone20.img.xz > /dev/sdX
```

where ubuntu-precise-12.04.2-armhf-3.8.13-bone20.img.xz will be replaced with the image file that you downloaded, and /dev/sdX will be replaced with your card ID, in this example /dev/sdb. Eject the card and you are ready to plug it into the board and boot.

Objective complete – mini debriefing

Make sure your BeagleBone Black is unplugged and install the micro SD card into the slot. Then apply power. After the boot, you should get a screen that looks like this:

```
Ubuntu 12.04.2 LTS   ubuntu-armhf tty1

ubuntu-armf login:
```

You can now log in to the system. You'll need to use the username and password of the image you downloaded (unfortunately, they are not the same for all images, but you should be able to easily find these in the same place you found your image). For my distribution the default username is `ubuntu` and the password is also `ubuntu`. Note that the password will not show up while you type it in. Remember this username and password, you'll need to use it throughout the examples in this book. Entering those will bring you to this state:

```
Ubuntu 12.04.2 LTS  ubuntu-armhf tty1

ubuntu-armf login: ubuntu
Welcome to Ubuntu 12.04.2 LTS (GNU/Linux 3.8.13-bone20 armv71)

* Documentation:  https:/help.ubuntu.com/

The programs included with the Ubuntu system are free software;
The exact distribution terms for each program are described in the
individual files in /usr/share/doc/*/copyright.

Ubuntu comes with ABSOLUTELY NO WARRANTY, to the extent permitted by
applicable law.

ubuntu@ubuntu-armhf:~$
```

You should now be logged in and ready to start issuing commands.

Classified intel

Now, two questions arise: do you need an external computer during the creation of your projects? and what sort of computer do you need? The answer to the first question is a resounding yes. Most of your projects are going to be self-contained robots with very limited communication capabilities. You will be using an external computer to issue commands and see what is going on inside your BeagleBone Black. The answer to the second is a bit more difficult. Because you are working in Linux, most notably Ubuntu on your BeagleBone Black, there are some advantages to having an Ubuntu system available as your remote system. You'll be able to try some things on your computer before trying them in your embedded system. You'll also be working with similar commands for both, which will help your learning curve.

However, bulk of the personal computers today run some sort of Windows operating system, so that is what will be normally available. You can do almost all that you need to do as far as issuing commands and displaying information with a Windows machine, so either way will work. I'll try to give examples for both, as long as it is practical.

There is one more choice, the choice I actually prefer. I have access to both systems on my PC. Previously this was done by a process called dual booting, where both systems were installed on the computer and the user chose which system they wanted to run during boot-up. Changing systems in this kind of configuration was time consuming, however, and it used up a lot of disk space. There is a better way.

On my Windows machine, I have a virtual Ubuntu machine running under a free program from Oracle called VirtualBox. This program lets me run a virtual Ubuntu machine hosted by my Windows operating system. That way I can try things in Ubuntu, yet keep all the functionality of my Windows machine. I'm not going to explain how to install this; there is plenty of help on the Web. Just search for Ubuntu and VirtualBox. There are several websites that offer easy, step-by-step instructions. One of my favorites is `http://www.psychocats.net/ubuntu/`.

Adding a graphical user interface

You now have your Ubuntu system up and working, and you can type in commands and see their result in the terminal window. However, you need to add some additional basic functionality before adding all the cool capabilities that will make your robots walk, talk, and interact. First, you need to connect to the Internet so that you can update your system and add additional functionality. Second, in many of your projects, you will be working with graphical programs, most notably when you connect webcams or other image sensors.

Prepare for lift off

You're going to need a graphical user interface (GUI), so let's tackle that problem.

Simply run a LAN cable from a router or switch to the BeagleBone board, plug it in the LAN connector, and restart the BeagleBone.

Now type `ifconfig` at the prompt. You should get a display like the following:

```
ubuntu@ubuntu-armhf:~$ ifconfig
eth0      Link encap:Ethernet  HWaddr c8:a0:30:bd:2c:9e
          inet addr:157.201.194.187  Bcast:157.201.194.255  Mask:255.255.255.128
          inet6 addr: fe80::caa0:30ff:febd:2c9e/64 Scope:Link
          UP BROADCAST RUNNING MULTICAST  MTU:1500  Metric:1
          RX packets:1813 errors:0 dropped:0 overruns:0 frame:0
          TX packets:152 errors:0 dropped:0 overruns:0 carrier:0
          collisions:0 txqueuelen:1000
          RX bytes:190064 (190.0 KB)  TX bytes:23358 (23.3 KB)
          Interrupt:56

lo        Link encap:Local Loopback
          inet addr:127.0.0.1  Mask:255.0.0.0
          inet6 addr: ::1/128 Scope:Host
          UP LOOPBACK RUNNING  MTU:65536  Metric:1
          RX packets:0 errors:0 dropped:0 overruns:0 frame:0
          TX packets:0 errors:0 dropped:0 overruns:0 carrier:0
          collisions:0 txqueuelen:0
          RX bytes:0 (0.0 B)  TX bytes:0 (0.0 B)

ubuntu@ubuntu-armhf:~$ █
```

This tells you that you are connected to the Internet and have a valid Internet address. In this case the valid address is 157.201.194.187. This address has been assigned by your Internet router.

Generally there are two types of IP addresses that your board can be assigned: one is called static and the other dynamic. In the static case you will always be assigned the same address. In the dynamic case, the address may change each time the system boots, as it asks the system for an address, which it then uses for that session. Most systems are configured for the dynamic case; however, if your system isn't changing, you will most likely get the same address each time you power on and log in to the system. To learn more about DHCP, try http://www.teracomtraining.com/tutorials/teracom-tutorial-dynamic-IP-addresses-and-DHCP.htm.

Once you get here, you'll want to update your operating system. Type in `sudo apt-get update`. The system will prompt you for the [sudo] password, which is the same as the password you have already been using. Once you enter this, the system will automatically go out and find all the updates associated with the system and applications that you have installed. This may take a very long time, depending on how out-of-date your system has become.

Engage thrusters

Now that you are connected to the Internet and have updated your Ubuntu system, you need to install a graphical user interface. Ubuntu generally comes with a very full-featured windowing system. However, it uses a good deal of memory and can interfere with the performance you may need later. So you are going to install a "light" Windows system on top of your Ubuntu distribution. There are several choices; I like to use Xfce. It is stable, seems to work well, and offers a fairly complete set of capabilities while not overwhelming your system resources. To install this, type `sudo apt-get install xfce4` in the command prompt. Again, the system will prompt you for your password and then start the install. This install will take a significant amount of time as it is installing not only the windowing system, but a number of packages the windowing system depends on.

Just a brief note about installing SW. You will be using `apt-get` to install SW throughout this book. This is the command that Ubuntu uses to go out and find SW and then install it on your machine. The nice thing about this is that it will also normally search and find dependencies and download them as well. Thus not only the package you want, but the packages that are needed for that package are installed as well. However, a bit of caution: this is not fool-proof! You will find times when the SW you have installed will not function because of a dependency that the system does not know about.

Objective complete – mini debriefing

Once the Xfce Windows system is installed, reboot your system by typing `sudo reboot`. The system will go down and then should come back to the log-in screen. Log in, then type `startx` at the prompt. After some time the Windows system will come alive. The first time you run the system you will get **Welcome** to the first start of the panel and a prompt, which will ask you which setup you want for the first setup. Choose the **User default config** selection.

Then you will see the following screen:

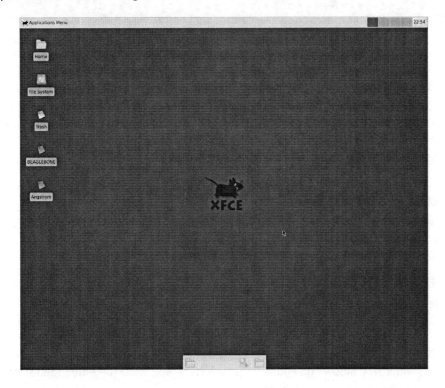

If you see the mouse, then you are successful!

Classified intel

You're probably asking yourself why you didn't copy your image to the internal eMMC memory card instead of just leaving your card in the system. There are two reasons, really. The first is space. The 2 GB that are available in the eMMC is not sufficient to build many, if not most, of the projects you're going to be working on, so you'll need a card anyway. Second, you will find occasions where you want to start over from scratch. Using a card makes this very easy; the eMMC process is a bit more permanent. There are several sites that can show you how to create your Ubuntu system on the eMMC internal memory space, but I'm not going to cover that here. Unfortunately, there are some downsides to not using the eMMC internal memory space, the system will boot slower and you have the additional cost of the external card, but in the long run it will be worth it.

Accessing the board remotely

You now have a very usable Ubuntu computer system. You can use it to access the Internet, write riveting novels, balance your accounts—just about anything you could do with a standard personal computer. However, that is not your purpose; you want to use your embedded system to power your delightfully inventive projects. In most cases you won't want to connect a keyboard, mouse, and display to your projects, as you will want to keep your robot sizes small and maneuverable. However, you still need to communicate with your device, program it, and have it tell you what is going on when things don't work right. You'll spend some time on this task establishing remote access to your device.

Prepare for lift off

To complete this task you'll need to have your external PC connected to the LAN, other than that, you are ready to go.

Engage thrusters

There are three ways you are going to access your system from your external PC:

▶   The first is through a simple terminal interface using the SSH protocol.

▶   The second way is using a program called vncserver, which will allow you to open a graphical "window" on your PC that will show you what the embedded system would be displaying on its display.

▶   Finally, if you are using Microsoft Windows on your remote computer, I'll show how you can transfer files via a program called WinScp, which is custom made for this purpose.

So, first, make sure your basic system is up and working. Open a terminal window and check the IP address of your unit. You're going to need this no matter how you want to communicate with the system. You do this by issuing the `ifconfig` command. You should get something that looks like this:

```
ubuntu@ubuntu-armhf:~$ ifconfig
eth0      Link encap:Ethernet  HWaddr c8:a0:30:bd:2c:9e
          inet addr:157.201.194.187  Bcast:157.201.194.255  Mask:255.255.255.128
          inet6 addr: fe80::caa0:30ff:febd:2c9e/64 Scope:Link
          UP BROADCAST RUNNING MULTICAST  MTU:1500  Metric:1
          RX packets:198 errors:0 dropped:0 overruns:0 frame:0
          TX packets:64 errors:0 dropped:0 overruns:0 carrier:0
          collisions:0 txqueuelen:1000
          RX bytes:20676 (20.6 KB)  TX bytes:8905 (8.9 KB)
          Interrupt:56

lo        Link encap:Local Loopback
          inet addr:127.0.0.1  Mask:255.0.0.0
          inet6 addr: ::1/128 Scope:Host
          UP LOOPBACK RUNNING  MTU:65536  Metric:1
          RX packets:0 errors:0 dropped:0 overruns:0 frame:0
          TX packets:0 errors:0 dropped:0 overruns:0 carrier:0
          collisions:0 txqueuelen:0
          RX bytes:0 (0.0 B)  TX bytes:0 (0.0 B)

ubuntu@ubuntu-armhf:~$
```

You'll need that "inet address" to contact your board via the LAN connection. First, let's configure an SSH terminal from your remote computer. An SSH terminal is a **Secure Shell Hypterminal** connection, which simply means you'll be able to access your board and type in commands at the prompt, just like you have done without the Windows system. In order to do this, you need to have an SSH type application on your remote computer. If you are running Microsoft Windows, you can download such an application. My personal favorite is PuTTY. It is free and does a very good job of allowing you to save your configuration so that you don't have to type it in each time. Type `putty` in a search window, and you'll soon come to a page that supports a download, or you can go to `www.putty.org`.

Download PuTTY to your Microsoft Windows machine. Then run PuTTY by going to the directory where it has been placed and double-clicking on the file `putty.exe`. You should see a configuration window. It will look something like this:

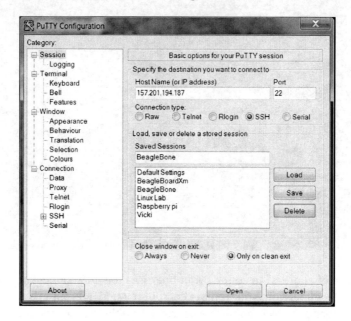

Type the inet address from the previous page in the **Host Name** space and make sure the SSH selection is highlighted. I save this configuration under BeagleBone so that I can load it each time.

When you press **Open**, the system will try to open a terminal window onto your BeagleBone Black via the LAN connection. The first time you do this, you will get a warning about an RSA key, as the two computers don't "know" about each other, so Windows is complaining that a computer that it doesn't know is about to be connected in a fairly intimate way. Simply select **OK**, and you should get a terminal with a login prompt.

Now you can log in and issue commands to your BeagleBone Black. If you'd like to do this from a Linux machine, the process is even simpler. Bring up a terminal window and then type `ssh ubuntu@157.201.194.187 -p 22`. This will then bring you to the login screen of your BeagleBone Black, which should look similar to the preceding screenshot.

SSH is a really useful tool to communicate with your BeagleBone Black, and I use it extensively. However, sometimes you need a graphical look at your system, and you don't necessarily want to connect a monitor or a small LCD display. You can get this by using an application called vncserver. First, let's install a version of this on your BeagleBone Black by typing `sudo apt-get install tightvncserver` in a terminal window on your BeagleBone Black. This is a perfect opportunity to use SSH, by the way.

Tightvncserver is an application that will allow you to remotely view your complete windows system. Once you have it installed, you'll need to start the server by typing `vncserver` in a terminal window on the BeagleBone black. You will then be prompted for a password as shown in the following screenshot:

This can, and should be a different password than your password to access your BeagleBone Black. This will be the password your remote system will send to access the vncserver running on the board. Select a password—you don't need to set the password for the view only capability—and then your vncserver will be running.

You'll need a VNC viewer application for your remote computer. On my Windows system I use an application called Real VNC. When I start it up it gives me the following screen:

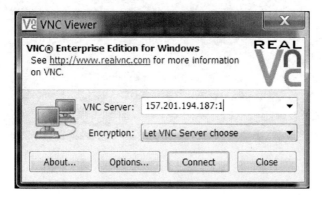

Enter the VNC Server address, which is the IP address of your BeagleBone Black, with a **:1** after it, and select **Connect**. You will get this pop up:

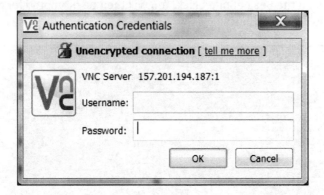

Type in the password you just entered while starting the vncserver, and you should then get a graphics view of your BeagleBone Black. Hopefully that looks like this:

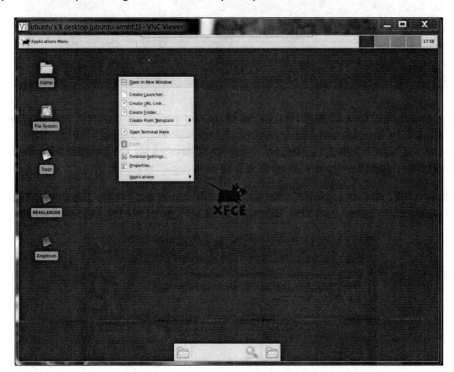

You can now access all the capabilities of your system, albeit they may be slower if you are doing graphics-intensive data transfers. You'll see this as you work through your projects.

There are ways to make your vncserver on your BeagleBone Black start automatically on boot. I have not used them; I choose to type the command `vncserver` from an SSH application when I want the application running. This keeps your running applications to a minimum and, more importantly, provides for fewer security risks. If you'd like to start your vncserver each time you boot, there are several places on the Internet that will show you how to configure this. Try `http://www.havetheknowhow.com/Configure-the-server/Run-VNC-on-boot.html`. You will only have to type the password the very first time you start vncserver—it will remember it after that. You also do not need to start Xfce using the `startx` command on your BeagleBone Black for it to come up in the VNC viewer to start its view onto the graphical interface.

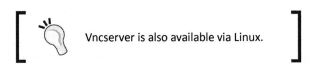

Vncserver is also available via Linux.

The final piece of SW I like to use with my Windows system is a free application called WinSCP. To download and install this piece of SW, simply search the web for WinSCP and follow the instructions. Once installed, run the program. It will open the following dialogue box:

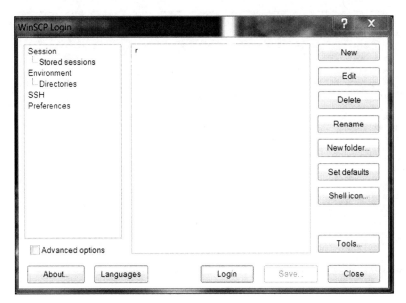

Select **New**, and you will get the following screen:

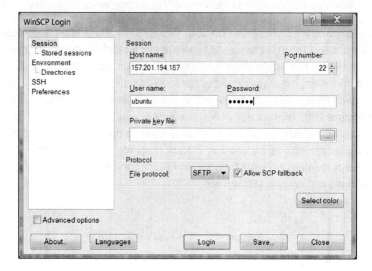

Here you fill in the IP address in the host name, `ubuntu` in the user name, and the BeagleBone Black password (not the vncserver password) in the password space. Press **log-in** and you should then see the application displayed:

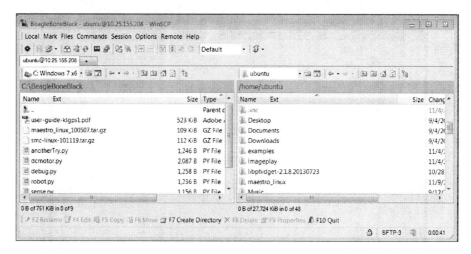

Now you can drag-and-drop files from one system to the other.

Objective complete – mini debriefing

Once you've completed this step, you can now access your system fully remotely without connecting a display, keyboard, and mouse. Now your system will look like this:

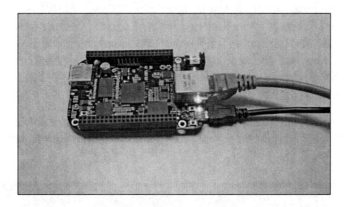

You only need to connect the power and LAN. If you need to issue simple commands, you'll connect via SSH. If you need a more complete set of graphical functionality, you can access this via vncserver. Finally, if you are using a Windows system and want to transfer files back and forth, you have access to WinScp. Now you have the toolkit you need to build your first capabilities.

Classified intel

One of the challenges of accessing the system remotely is that you need to know the IP address of your board. If you have the board connected to a keyboard and display, you can always just run `ifconfig` to get this info. But you're going to use the board in many applications where you don't have this information. There is a way to discover this by using an ipscanner application. There are several available for free; simply download and use as they instruct. They can let you know what addresses are being used, and this should then let you find your BeagleBone Black address without typing `ipconfig`. From Linux you can use an application called nmap.

Mission accomplished

Congratulations! You've completed the first stage of your journey. You have your BeagleBone Black up and working. No gathering dust in the bin for this piece of HW. It is now ready to start connecting to all sorts of interesting devices in all sorts of interesting ways. One thing that is important to note: the system is not going to be nearly as stable as your PC or Mac. The HW and SW is new, and you're going to find yourselves coming back to this project from time to time to revive your board when it dies, because your SW has put it in some unrecoverable state.

Generally it is difficult to physically harm the board as long as you are only using the USB interface, so be brave, and remember, you often learn more from your failures than your successes.

A challenge

Your system has lots of capabilities. Feel free to play with the system—try to get an understanding of what is already there and what you'll want to add from a SW perspective. One advanced possibility is to connect the BeagleBone Black via a wireless LAN connection, so that you don't have to connect a HW connection when you want to communicate with it. There are several good tutorials on the Internet. Here is a place to start, `http://elinux.org/Beagleboard:BeagleBoneBlack`. If you have the desire, feel free to follow them to see if you can get a wireless LAN connection working.

Remember, there is limited power on your USB port, so make sure you have a powered USB hub before having a go. Usually you'll need a powered USB hub that can supply power greater than 1 amp.

2
Programming the BeagleBone Black

Before you get started with building your robotic projects, let's take a bit of time to either introduce or review how to program the BeagleBone Black.

Mission briefing

Now that you are up and running, you'll want your BeagleBone Black to start doing something. This requires you to either create your own programs, or edit an existing program. This chapter will provide a brief introduction into editing a file and programming.

Why is it awesome?

It is fun to build hardware, and you'll spend a good deal of time designing and building your robots, but without programming, your robots won't get very far. This chapter will introduce you to file editing and programming concepts, so you'll feel comfortable creating some of the fairly simple programs that we'll talk about through the book. You'll also know how to change programs that are already available, making your robot do even more amazing things.

Your objectives

In this chapter we will:

- ▸ Introduce some of the basic Linux commands and show how to navigate around the filesystem on the BeagleBone Black
- ▸ Show how to create, edit, and save files on the BeagleBone Black
- ▸ Learn how to create and run Python programs on the BeagleBone Black

- ▸ Introduce some of the basic programming constructs on the BeagleBone Black
- ▸ Show how the C++ programming language is both similar and different so you can understand when you need to change C++ code files

Mission checklist

We're going to use the basic configuration that you created in *Chapter 1, Getting Started with the BeagleBone Black*. You can either accomplish the tasks in this chapter by connecting a keyboard, mouse, and monitor to the BeagleBone Black or by remotely logging into the BeagleBone Black using vncserver, or remotely logging using SSH. Any of these methods will work to complete the examples in this chapter.

Basic Linux commands and navigating the filesystem

After completing *Chapter 1, Getting Started with the BeagleBone Black*, you should have a working BeagleBone Black running a version of Linux called Ubuntu. We selected this distribution because it is the most popular and has the largest set of supported hardware and software. The commands I am going to review should also work with other versions of Linux, but I'll be showing examples using Ubuntu.

Prepare for lift off

So, power up your BeagleBone Black and log in using the proper username and password. If you are going to log in remotely, go ahead and establish the connection and log in. Now we will take a quick tour of Linux. This will not be extensive, but we will walk through some of the basic commands.

Engage thrusters

Once you have logged in, you should open up a terminal window. If you are logging in using a keyboard, mouse, and monitor, or using vncserver, you'll find the terminal selection by selecting the **Applications Menu** in the upper-left hand corner, then selecting the **Terminal Emulator** selection, as shown in the following screenshot:

Downloading the example code and colored images

You can download the example code files and colored images for this Packt book you have purchased from your account at http://www.packtpub.com. If you purchased this book elsewhere, you can visit http://www.packtpub.com/support and register to have the files e-mailed directly to you.

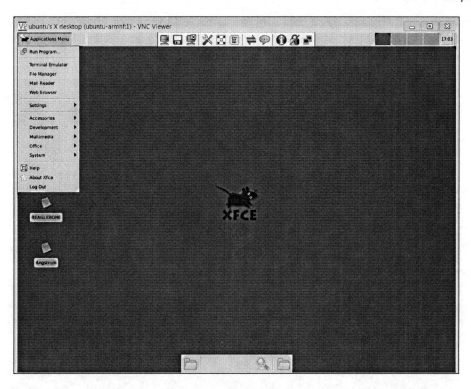

If you are using PuTTY to use SSH to log in, you should already be at the terminal emulator program. Either way, it should look something like this:

Your cursor is at the command prompt. Unlike Microsoft Windows, or Apple's OS, most of our work will be done by actually typing commands into the command line. So, let's try a few. First, type `ll`, and you should something like this:

```
ubuntu@ubuntu-armhf: ~
-rw-------  1 ubuntu ubuntu  106 Oct 29 17:02 .Xauthority
-rw-------  1 ubuntu ubuntu  128 Oct 29 17:06 .bash_history
-rw-r--r--  1 ubuntu ubuntu  220 Apr  3  2012 .bash_logout
-rw-r--r--  1 ubuntu ubuntu 3486 Apr  3  2012 .bashrc
drwx------  3 ubuntu ubuntu 4096 Oct 29 17:02 .cache/
drwx------  5 ubuntu ubuntu 4096 Oct 29 17:02 .config/
drwx------  3 ubuntu ubuntu 4096 Oct 29 17:02 .dbus/
-rw-r--r--  1 ubuntu ubuntu 3764 May  8 00:23 .dircolors
drwxr-xr-x  2 ubuntu ubuntu 4096 Oct 29 17:02 .fontconfig/
drwxrwxr-x  2 ubuntu ubuntu 4096 Oct 29 17:02 .gstreamer-0.10/
drwx------  2 ubuntu ubuntu 4096 Oct 29 17:02 .gvfs/
drwxrwxr-x  3 ubuntu ubuntu 4096 Oct 29 17:02 .local/
-rw-r--r--  1 ubuntu ubuntu  675 Apr  3  2012 .profile
drwx------  2 ubuntu ubuntu 4096 Oct 29 17:02 .vnc/
-rw-------  1 ubuntu ubuntu 2784 Oct 29 17:08 .xsession-errors
drwxr-xr-x  2 ubuntu ubuntu 4096 Oct 29 17:02 Desktop/
drwxr-xr-x  2 ubuntu ubuntu 4096 Oct 29 17:02 Documents/
drwxr-xr-x  2 ubuntu ubuntu 4096 Oct 29 17:02 Downloads/
drwxr-xr-x  2 ubuntu ubuntu 4096 Oct 29 17:02 Music/
drwxr-xr-x  2 ubuntu ubuntu 4096 Oct 29 17:02 Pictures/
drwxr-xr-x  2 ubuntu ubuntu 4096 Oct 29 17:02 Public/
drwxr-xr-x  2 ubuntu ubuntu 4096 Oct 29 17:02 Templates/
drwxr-xr-x  2 ubuntu ubuntu 4096 Oct 29 17:02 Videos/
ubuntu@ubuntu-armhf:~$
```

The command `ll` in Linux is an abbreviation for list-long and lists all the files and directories in our current directory with information about who owns the files, the times they were created, and the various permissions on the files. The files are listed by their names; you can tell the directories because they are normally in a different color, and `d` proceeds the lines for those listing. In this case `Videos` is a directory. The default installation of Ubuntu has no directories. Installing the Xfce windows manager creates the `Desktop`, `Documents`, `Downloads`, `Music`, `Pictures`, `Public`, `Templates`, and `Videos` directories.

You can move around the directory structure by issuing the `cd` (change-directory) command. For example, if you want to see what is in the `Videos` directory, type `cd ./Videos`. Now if you issue the `ll` command, you should see something like this:

This directory is empty, except for a couple of default directory selections. Now, I should point out that you used a shortcut when you typed `cd ./Videos`. The `.` is a shortcut for the default directory. You could also have typed `cd /home/ubuntu/Videos` and gotten the exact same result, because you were in the `/home/ubuntu` directory, which is the directory where you always start when you first log in to the system.

If you ever want to see which directory you are in, simply type `pwd`, which stands for print-working-directory. If you do that here, you should get:

The result is `/home/ubuntu/Videos`. Now, you can use two different shortcuts to move back to the default directory. The first is to type `cd ..`; this will take you to the directory just above this one in the hierarchy. Do this, then type `pwd`, and you should see the following:

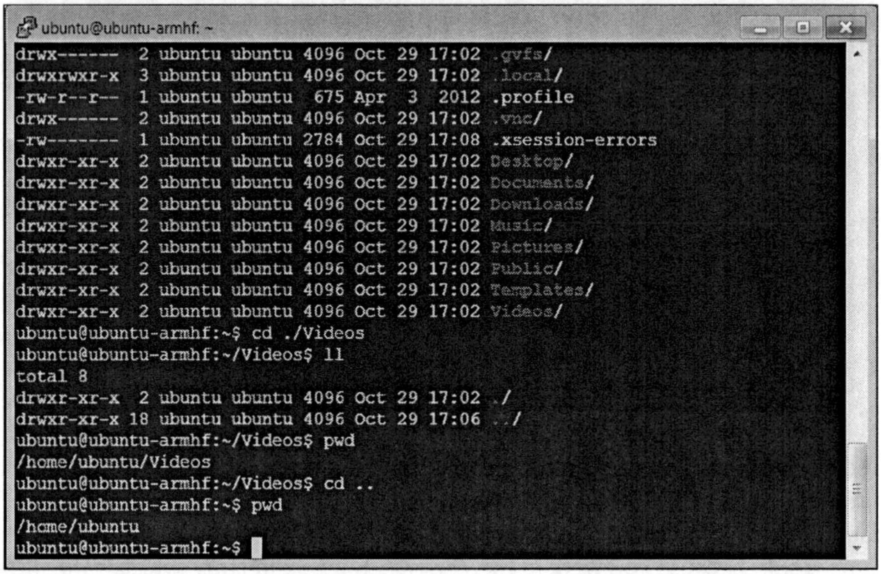

The other way to get back to the home directory is to type `cd ~`, as this will always return you to the home directory. If you were to do this from the `Videos` directory, and then type `pwd`, you will see something like this:

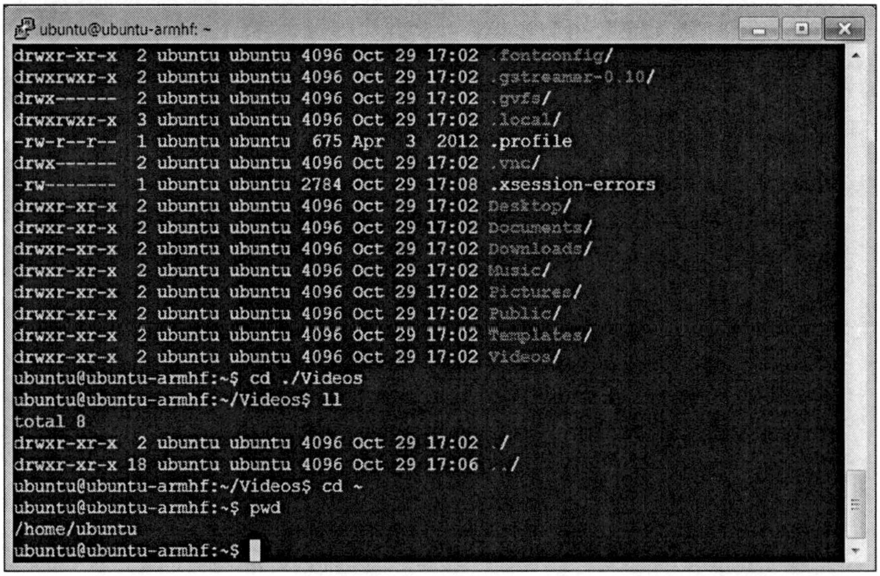

You can also use `cd -`, which will direct you to the last directory accessed. Another way to go to a specific file is to use the entire path name. In this case, if you want to go to the `/home/ubuntu/Video` directory from anywhere in the filesystem, simply type `cd /home/ubuntu/Video` and you will go to that directory.

There are a number of other Linux commands that you might find useful as you program your robot. Here is a table with some of the more useful commands:

Linux command	What it does
`ll`	List-long: Lists all the files and directories in the current directory. This includes lots of extra information about the file, including time it was created, permissions, owners, and so on.
`ls`	List-short: Lists all the files and directories in the current directory by just their names.
`rm filename`	Remove: Removes whichever file is specified by the filename.
`mv filename1 filename2`	Move: Renames filename1 to filename2.
`cp filename1 filename2`	Copy: Copies filename1 to filename2.
`mkdir directoryname`	Make directory: Make a directory with the name. `directoryname`. This will be made in the current directory unless otherwise specified.
`cat filename`	Catalog filename: Displays the file, you may want to use the `| less` command at the end of this so that it will display the file a page at a time. Use the space bar to go to the next page.
`clear`	Clear: Clears the current terminal window.
`sudo`	Super user: If you type the `sudo` command in front of any command, it will do that command as the super user. This can be required if the command or program you are trying to execute needs super user permissions. If, at any point in this book, you type a command or the name of program you want to run and it seems to suggest that the command does not exist, or permission is denied, try it again with sudo in front of the command or name of the program.

Objective complete – mini debriefing

Now you can play around and look at your system and the files that are available to you. Be a bit careful! Linux is not like Windows; it will not warn you if you try to delete a file, or copy over a current file.

Creating, editing, and saving files on the BeagleBone Black

Now that you can log in and move easily between directories and see which files are in your directories, you'll want to be able to edit those files. To do this, you'll need a program that allows you to edit the characters in a file. If you are used to working in Microsoft Windows, you probably have used a program like Microsoft Notepad, Wordpad, or Word to do this. As you might imagine, these are not available in Linux. There are several choices, all of which are free. I am going to show you how to use an editor program called **Emacs**. Other possibilities are programs like nano, vi, vim, and gedit. Programmers have strong preferences about which editor to use, so if you already have a favorite, you can skip this section.

Prepare for lift off

If you want to use Emacs, then download and install Emacs by typing `sudo apt-get install emacs`.

Engage thrusters

Once installed, you can run Emacs simply by typing `emacs filename`, where filename is the name of the file you want to edit. If the file does not exist, then Emacs will create the file. Here is what you will see if you type `emacs example.py` at the prompt:

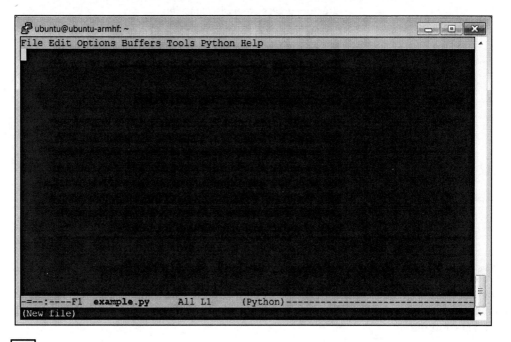

Notice that unlike Windows, Linux doesn't automatically assign file extensions; it is up to us to specify what kind of file we want to create. Notice that Emacs has indicated, in the lower left, that you have opened a new file. Now, if you are using Emacs in the Xfce windows interface, either because you have a monitor, keyboard, and mouse hooked up or are running vncserver, you can use the mouse in much the same way that you use the mouse in the Microsoft world.

However, if you are running Emacs from SSH, you won't have the mouse available, so you'll need to move around the file using the cursor keys. You'll also have to use some keystroke commands to save your file, as well as accomplish a number of other tasks you would normally use the mouse to select. For example, when you are ready to save the file you'll use *Ctrl + X* then *Ctrl +S*, and that will save the file under the current filename. When you want to quit Emacs you'll use *Ctrl + X* then *Ctrl + C*. This will stop Emacs and return you to the command prompt. If you are going to use Emacs inside of Emacs, here are a number of keystroke commands you might find useful:

emacs command:	What it does
Ctrl + X and *Ctrl + S*	Save: Saves the current file.
Ctrl + X and *Ctrl + C*	Quit: Exits Emacs and returns to the command prompt.
Ctrl + K	Kill: Erases the current line.
Ctrl + U	Undo: Reverts the last action.
Left mouse button: text selection Cursor: right mouse button	Cut and paste: If you select the text you want to paste with the mouse using the left mouse button, then move the cursor to where you want to paste the code, then hit the right mouse button, the code will be pasted to that location.

Objective complete – mini debriefing

Now that you have the capability to edit files, in the next section you'll use this capability to create programs.

Creating and running Python programs on the BeagleBone Black

Now that you can get around, and even edit programs, you can begin to use the BeagleBone Black to create programs so you can control your robotic projects.

Prepare for lift off

Now that you are ready to begin programming, you'll need to choose a language. There are many available, C, C++, Java, Python, Perl, and a great deal of other possibilities. I'm going to introduce you to Python for two reasons. First, it is a straightforward language that is intuitive and very easy to use. Second, much of the open source functionality in the robotics world is available in Python. We'll also cover a bit of C in this chapter as well, as some functionality is only available in C. But it makes most sense to start in Python. To work the examples in this section, you'll need a version of Python installed to complete this section. Fortunately the basic Ubuntu system has a version already installed, so you are ready to begin.

We are going to just cover some of the very basic concepts here. If you are new to programming, there are a number of different websites that provide interactive tutorials. If you'd like to practice some of the basic programming concepts in Python using these tools, try `www.codeacademy.com` or `http://www.learnpython.org/` and give it a try. There are also a number of excellent books, for example, *A Byte of Python*.

Engage thrusters

In this section we'll cover how to create and run a Python file. It turns out that Python is an interactive language, so you could run Python and then type in commands one at a time. But we want to use Python to create programs, so we are going to type our commands using Emacs and then run them from the command line by invoking Python. Let's get started.

Open an example Python file by typing `emacs example.py`. Now, let's put some code in the file. Start with these five lines:

Note, your code may be color coded. I have removed the color coding here so that it is easier to read.

Here is an explanation of the code.

Let's go through the code to see what is happening:

1. `a = input("Input value: ")`: One of the basic needs of a program is to get input from the user. `raw_input` allows us to do that. The data will be input by the user and stored in the variable `a`. The prompt `Input value:` will be shown to the user.

2. `b = input("Input second value: ")`: This data will also be input by the user and stored in the variable `b`. The prompt `Input second value:` will be shown to the user.

3. `c = a + b`: This is an example of something you can do with the data; in this example you can add the variables `a` and `b`.

4. `print c`: Another basic need of our program is to print out results. The print command prints out the value of `c` to the display.

Once you have created your program, save it (using *Ctrl + X* then *Ctrl + sS*) and quit Emacs (using *Ctrl + X* then *Ctrl + C*). Now from the command line run your program by typing `python example.py`. You should see something like this:

```
ubuntu@ubuntu-armhf: ~
ubuntu@ubuntu-armhf:~$ emacs example.py
ubuntu@ubuntu-armhf:~$ python example.py
Input value: 4
Input second value: 5
9
ubuntu@ubuntu-armhf:~$
```

You can also run the program right from the command line without typing `python filename` by adding one line to the program. Now the program looks like this:

```
ubuntu@ubuntu-armhf: ~
File Edit Options Buffers Tools Python Help
#!/usr/bin/python

a = input("Input value: ")
b = input("Input second value: ")
c = a + b
print c




-=--:----F1   example.py      All L3        (Python)---------------------------
Wrote /home/ubuntu/example.py
```

Adding `#!/usr/bin/python` as the first line simply makes this file available for us to execute from the command line. Once you have saved the file and exited Emacs, type `chmod +x example.py`. This will change the file's execution permissions so the computer will now believe it and execute it. You should be able to simply type `./example.py` and the program should run, like this:

```
ubuntu@ubuntu-armhf: ~
ubuntu@ubuntu-armhf:~$ chmod +x example.py
ubuntu@ubuntu-armhf:~$ ls
#try1.py#  Documents  Music      Public     Videos        example.py~
Desktop    Downloads  Pictures   Templates  example.py
ubuntu@ubuntu-armhf:~$ example.py
-bash: example.py: command not found
ubuntu@ubuntu-armhf:~$ ./example.py
Input value: 6
Input second value: 7
13
ubuntu@ubuntu-armhf:~$
```

Notice that if you simply type `example.py`, the system will not find the executable file. In this case the file has not been registered with the system, so you have to give it a path to the file, in this case `./` is the current directory.

Objective complete – mini debriefing

Now that you know how to create, enter, and run your simple Python programs, let's look at some programming constructs.

Basic programming constructs on the BeagleBone Black

Now that you know how to enter and run a simple Python program on the BeagleBone Black, let's look at some more complex programming constructs. Specifically, we'll cover what to do when we want to decide which instructions to execute and show how to loop our code to do the same thing more than once. I'll give a brief introduction into how to use libraries in the Python code, and how to organize statements into functions. Finally I'll very briefly cover object oriented code organization.

Prepare for lift off

As with the previous section, once you have the basic system and Emacs, you are ready to start coding.

Engage thrusters

As you have seen, your programs normally start with the first line of code and then continue, executing the next line, until your program runs out of code. This is fine, but what if you want to decide between two different courses of action? We can do this in Python using an `if` statement. Here is some example code:

```
ubuntu@ubuntu-armhf: ~
File Edit Options Buffers Tools Python Help
#!/usr/bin/python

a = input("Input value: ")
b = input("Input second value: ")
if a > b:
    c = a - b
else:
    c = b - a
print c



-=--:----F1   example.py      All L8      (Python)------------------------------
Wrote /home/ubuntu/example.py
```

Here is the detail, line by line:

1. `#!/usr/bin/python`: This is included so that you can make your program executable.

2. `a = input("Input value: ")`: One of the basic needs of a program is to get input from the user. `raw_input` allows us to do that. The data will be input by the user and stored in the variable a. The prompt `Input value:` will be shown to the user.

3. `b = input("Input second value: ")`: This data will also be input by the user and stored in the variable b. The prompt `Input second value:` will be shown to the user.

4. `if a > b:`: This is an `if` statement. The expression is evaluated, in this case `a > b`. If it is true, the program will do the next statement(s) that are indented. If not, it will skip those statement(s). In this case `c = a - b`.

5. `else:`: The `else` is an optional part of the command. If the expression in the `if` statement is evaluated as false, then the indented statement(s) will be executed, in this case `c = b - a`.

6. `print c`: Another basic need of our program is to print out results. The `print` commands prints out the value of c to the display.

You can run this program a couple of times, checking both possibilities of the `if` statement:

```
ubuntu@ubuntu-armhf: ~
ubuntu@ubuntu-armhf:~$ ./example.py
Input value: 4
Input second value: 3
1
ubuntu@ubuntu-armhf:~$ ./example.py
Input value: 3
Input second value: 5
2
ubuntu@ubuntu-armhf:~$
```

Another useful construct is the `while` construct; it will allow us to execute a set of statements over and over until a specific condition has been met. Here is a piece of code that uses this construct:

```
ubuntu@ubuntu-armhf: ~
File Edit Options Buffers Tools Python Help
#!/usr/bin/python

a = 0
b = 1
while a != b:
    a = input("Input value: ")
    b = input("Input second value: ")
    c = a + b
    print c




-=--:----F1  example.py     All L1      (Python)------------------------------
For information about GNU Emacs and the GNU system, type C-h C-a.
```

Here are the details of this code:

1. `#!/usr/bin/python`: This is included so you can make your program executable.

2. `a = 0`: Set the value of variable a to 0. We'll need this only to make sure we do the loop at least once.

3. `b = 1`: Set the value of variable b to 1. We'll need this only to make sure we do the loop at least once.

4. `while a != b:`: The expression a `!=` b (in this case `!=` means not equal to) is checked. If it is true, then the statement(s) that are indented are executed. When the statement evaluates as false, then the program jumps to the statements after the indented section.

5. `a = input("Input value: ")`: One of the basic needs of a program is to get input from the user. `raw_input` allows us to do that. The data will be input by the user and stored in the variable a. The prompt `Input value:` will be shown to the user.

6. `b = input("Input second value: ")`: This data will also be input by the user and stored in the variable b. The prompt `Input second value:` will be shown to the user.

7. `c = a + b`: The variable c is loaded with the sum of a and b.

8. `print c`: The `print` commands prints out the value of c to the display.

Now you can run the program, and notice that when you enter the same value for a and b, the program stops as shown in the following screenshot:

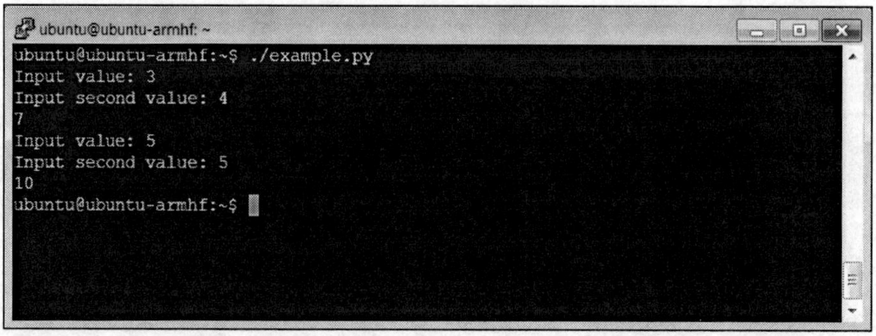

```
ubuntu@ubuntu-armhf: ~
ubuntu@ubuntu-armhf:~$ ./example.py
Input value: 3
Input second value: 4
7
Input value: 5
Input second value: 5
10
ubuntu@ubuntu-armhf:~$
```

The next concept we need to cover is how to put a set of statements into a function. Here is the code:

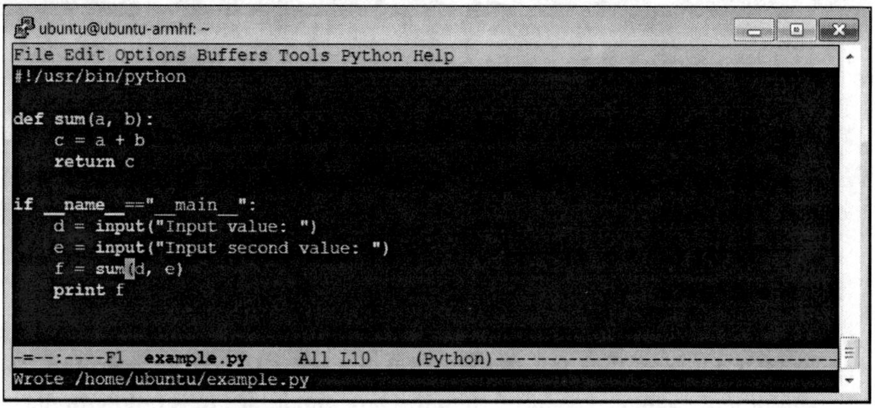

```
ubuntu@ubuntu-armhf: ~
File Edit Options Buffers Tools Python Help
#!/usr/bin/python

def sum(a, b):
    c = a + b
    return c

if __name__ == "__main__":
    d = input("Input value: ")
    e = input("Input second value: ")
    f = sum(d, e)
    print f

-=--:----F1  example.py      All L10     (Python)--------------------
Wrote /home/ubuntu/example.py
```

And here is the explanation of the code:

1. `#!/usr/bin/python`: This is included so that you can make your program executable.

2. `def sum(a, b):`: This defines a function whose name is `sum`. This function takes to arguments, `a` and `b`.

3. `c = a + b`: Anytime the `sum` function is called, it will add the value in `a` with the value in `b`.

4. `return c`: When the function is finished it will return `c` to the calling expression.

5. `if __name__ == "__main__":`: In this particular case, you don't want your program to start at the top of the file and then execute each statement, rather you want to start here. This line tells the program to begin its execution at this point.

6. `d = input("Input value: ")`: This data will also be input by the user and stored in the variable d. The prompt `Input value:` will be shown to the user.

7. `e = input("Input second value: ")`: This data will also be input by the user and stored in the variable e. The prompt `Input second value:` will be shown to the user.

8. `f = sum(d, e)`: The function `sum` is called. The value in variable d is copied into the variable a in the `sum` function, and the variable e is copied to the variable b in the `sum` function. The program then goes to the `sum` function and executes it. The return value is then stored in the variable `f`.

9. `print f`: The `print` commands prints out the value of `f` to the display.

And here is the result when you run the code:

```
ubuntu@ubuntu-armhf: ~
ubuntu@ubuntu-armhf:~$ ./example.py
Input value: 4
Input second value: 2
6
ubuntu@ubuntu-armhf:~$ 
```

The next topic we need to cover is how to add functionality to our programs using libraries. Libraries include functionality that someone else has created that you want to add to your code. As long as the functionality exists, and your system knows about it, then you can include the library. So let's modify our code again:

```
ubuntu@ubuntu-armhf: ~
File Edit Options Buffers Tools Python Help
#!/usr/bin/python

import time

if __name__=="__main__":
    d = input("Input value: ")
    time.sleep(1)
    e = input("Input second value: ")
    f = d + e
    print f

-=--:----F1  example.py    All L4    (Python)----------------------
Wrote /home/ubuntu/example.py
```

And here is the line-by-line description of the code:

1. `#!/usr/bin/python`: This is included so that you can make your program executable.

2. `import time`: This includes the time library. The time library includes a function that allows you to pause for a certain number of seconds.

3. `if __name__=="__main__":`: In this particular case, you don't want your program to start at the top of the file and then execute each statement, rather you want to start here. This line tells the program to begin its execution at this point.

4. `d = input("Input value: ")`: This data will also be input by the user and stored in the variable d. The prompt `Input value:` will be shown to the user.

5. `time.sleep(1)`: This line calls the sleep function in the time library, which will cause a 1 second delay.

6. `e = input("Input second value: ")`: This data will also be input by the user and stored in the variable e. The prompt `Input second value:` will be shown to the user.

7. `f = d + e`: The variable f is loaded with the value of d + e.

8. `print f`: The `print` commands prints out the value of f to the display.

And your result:

```
ubuntu@ubuntu-armhf: ~
ubuntu@ubuntu-armhf:~$ ./example.py
Input value: 3
Input second value: 2
5
ubuntu@ubuntu-armhf:~$
```

Of course this looks very similar to the other results. But, you will notice a pause when you enter the first value and the second value.

The final topic we need to cover is object oriented organization in your code. In object oriented programming we organize a set of related functions into an object. If, for example, we have a set of functions that are all related, you can place them in the same class and then call them by associating them with a specific class. This is a complex and difficult topic, but let me just show a simple example:

```
ubuntu@ubuntu-armhf: ~
File Edit Options Buffers Tools Python Help
#!/usr/bin/python

class ExampleClass(object):
    def add(self, a, b):
        c = a + b
        return c

if __name__=="__main__":
    example = ExampleClass()
    d = input("Input value: ")
    e = input("Input second value: ")
    f = example.add(d, e)
    print f

-=--:----F1  example.py    All L1    (Python)------------------------
Wrote /home/ubuntu/example.py
```

And here is an explanation of the code:

1. `#!/usr/bin/python`: This is included so that you can make your program executable.

2. `class ExampleClass(object):`: This defines a class named `ExampleClass`. This class can have any number of functions associated with it.

3. `def add(self, a, b):`: This defines the function `add` as part of the `ExampleClass`. We can have functions that have the same names as long as they belong to different classes. This function takes two arguments, `a` and `b`.

4. `c = a + b`: The statement is a simple adding of two values.

5. `return c`: The function returns the result of the addition.

6. `if __name__=="__main__":`: In this particular case, you don't want your program to start at the top of the file and then execute each statement, rather you want to start here. This line tells the program to begin its execution at this point.

7. `example = ExampleClass()`: This defines a variable named `example`, whose type is `ExampleClass`. It now has access to all the functions and variables associated with the class `ExampleClass`.

8. `d = input("Input value: ")`: This data will also be input by the user and stored in the variable d. The prompt `Input value:` will be shown to the user.

9. `e = input("Input second value: ")`: This data will also be input by the user and stored in the variable e. The prompt `Input second value:` will be shown to the user.

10. `f = example.add(d,e)`: The instance of `ExampleClass` is called, and its function `add` is executed by sending d and e to the function. The result is returned and stored in the variable f.

11. `print f`: The `print` commands prints out the value of f to the display.

And the result:

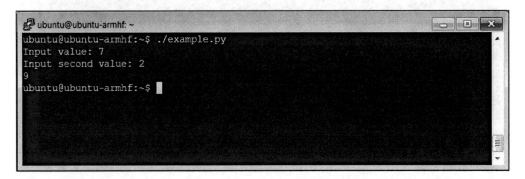

This result is the same as the others, there is no functionality difference; however, object-oriented techniques are used to keep like functions organized together to make code easier to maintain. It also makes it easier for others to use your code when the program size grows.

Objective complete – mini debriefing

Now you have a feel for the basics of Python coding, I'll introduce you very briefly to the C++ coding language.

Introduction to the C++ programming language

Now that you've been introduced to a simple programming language in Python, we need to spend a bit of time talking about a more complex but powerful language called C++. C++ is the original language of Linux, and has been around for many decades, but is still widely used by open source developers. It is similar to Python, but is also a bit different, and since you may need to understand and make changes to C++ code, you should be familiar with it and how it is used.

Prepare for lift off

As with Python, you will need to have access to the language capabilities. These come in the form of a compiler and build system, which turns your text files that contain programs to machine code that the processor can actually execute. To do this, type `sudo apt-get install build-essential`. This will install the programs you need to turn your code into executables for the system.

Engage thrusters

Now that the tools are installed, let's walk through some simple examples. Here is the first C++ code example:

```
ubuntu@ubuntu-armhf: ~
File Edit Options Buffers Tools C++ Help

#include <iostream>

int main()
{
  int a;
  int b;
  int c;

  std::cout << "Input value: ";
  std::cin >> a;
  std::cout << "Input second value: ";
  std::cin >> b;

  c = a + b;

  std::cout << c << std::endl;

  return 0;
}
-=--:----F1   example2.cpp    All L1      (C++/1 Abbrev)-----------------------
Wrote /home/ubuntu/example2.cpp
```

And here is an explanation of the code:

1. `#include <iostream>`: This is a library that is included so that your program can input data from the keyboard and output information to the screen.

2. `int main()`: As with Python, we can put functions and classes in the file, but you will always want to start execution at a known point; C++ defines this as the main function.

3. `int a;`: This defines a variable named `a`, of type `int`. C++ is a strongly typed language, which means we need to declare the type of the variable we are defining. The normal types are `int`: a number that has no decimal points, `float`: a number that requires decimal points, `char`: a character of text, and `bool`: a true or false value.

4. `int b;`: This defines a variable named `b`, of type `int`.

5. `int c;`: This defines a variable named `c`, of type `int`.

6. `std::cout << "Input value: ";`: This will display the string `Input value:` to the screen.

7. `std::cin >> a;`: The input that the user types will go into the variable `a`.

8. `std::cout << "Input second value: ";`: This will display the string `Input second value:` to the screen.

9. `std::cin >> b;`: The input that the user types will go into the variable `a`.

10. `c = a + b`: The statement is a simple adding of two values, placing the result into the variable `c`.

11. `std::cout << c << std::endl;`: The `cout` command prints out the value of `c` to the display. The `endl` at the end prints out a carriage return so that the next character appears on the next line.

12. `return 0;`: The `main` function ends and returns `0`.

To run this program, you'll need to run a compile process to turn it into an executable program that you can run. To do this, after you have created the program, type `g++ example2.cpp -o example2`. This will then process your program, turning it into a file that the computer can execute. The name of the executable program will be `example2` (as specified by the name after the output `-o` option).

If you do an `ll` on your directory after you have compiled this, you should see the `example2` file in your directory:

```
ubuntu@ubuntu-armhf: ~
-rw-rw-r--   1 ubuntu ubuntu   553 Oct 29 19:03 .emacs
drwx------   3 ubuntu ubuntu  4096 Oct 29 18:27 .emacs.d/
drwxr-xr-x   2 ubuntu ubuntu  4096 Oct 29 17:02 .fontconfig/
drwx------   2 ubuntu ubuntu  4096 Oct 29 18:28 .gconf/
drwxrwxr-x   2 ubuntu ubuntu  4096 Oct 29 17:02 .gstreamer-0.10/
drwx------   2 ubuntu ubuntu  4096 Oct 29 17:02 .gvfs/
drwxrwxr-x   3 ubuntu ubuntu  4096 Oct 29 17:02 .local/
-rw-r--r--   1 ubuntu ubuntu   675 Apr  3  2012 .profile
drwx------   2 ubuntu ubuntu  4096 Oct 29 17:02 .vnc/
-rw-------   1 ubuntu ubuntu  2784 Oct 29 17:08 .xsession-errors
drwxr-xr-x   2 ubuntu ubuntu  4096 Oct 29 17:02 Desktop/
drwxr-xr-x   2 ubuntu ubuntu  4096 Oct 29 17:02 Documents/
drwxr-xr-x   2 ubuntu ubuntu  4096 Oct 29 17:02 Downloads/
drwxr-xr-x   2 ubuntu ubuntu  4096 Oct 29 17:02 Music/
drwxr-xr-x   2 ubuntu ubuntu  4096 Oct 29 17:02 Pictures/
drwxr-xr-x   2 ubuntu ubuntu  4096 Oct 29 17:02 Public/
drwxr-xr-x   2 ubuntu ubuntu  4096 Oct 29 17:02 Templates/
drwxr-xr-x   2 ubuntu ubuntu  4096 Oct 29 17:02 Videos/
-rwxrwxr-x   1 ubuntu ubuntu   269 Oct 30 15:16 example.py*
-rwxrwxr-x   1 ubuntu ubuntu   269 Oct 30 15:16 example.py~*
-rwxrwxr-x   1 ubuntu ubuntu  8634 Oct 30 17:54 example2*
-rw-rw-r--   1 ubuntu ubuntu   230 Oct 30 17:52 example2.cpp
-rw-rw-r--   1 ubuntu ubuntu   229 Oct 30 17:52 example2.cpp~
ubuntu@ubuntu-armhf:~$
```

By the way, if you run into a problem, the compiler will try to help you figure out the problem. If, for example, you were to forget the `int` before `int  a;` you would get the following error when you try to compile:

```
ubuntu@ubuntu-armhf: ~
ubuntu@ubuntu-armhf:~$ g++ example2.cpp -o example2
example2.cpp: In function 'int main()':
example2.cpp:6:3: error: 'a' was not declared in this scope
ubuntu@ubuntu-armhf:~$
```

The error message indicates a problem in the `int main()` function, and tells you that the variable `a` was not successfully declared. Once you have the file compiled, to run the executable, type `./example2` and you should be able to create the following result:

```
ubuntu@ubuntu-armhf: ~
ubuntu@ubuntu-armhf:~$ g++ example2.cpp -o example2
ubuntu@ubuntu-armhf:~$ ./example2
Input value: 5
Input second value: 9
14
ubuntu@ubuntu-armhf:~$ 
```

I will not repeat the entire Python tutorial for C++ here; there are several good tutorials out on the Internet that can help. For example, at `http://www.cprogramming.com/tutorial/c-tutorial.html` and `http://thenewboston.org/list.php?cat=14`. There is one more aspect of C++ you will need to know about. The compile process that you just encountered seemed fairly straightforward. However, if you have your functionality distributed between a lot of files, or need lots of libraries, the command-line approach to executing a compile can get unwieldy.

The C++ development environment provides a way to automate this process, it is called the make process. When using this, you create a text program named `makefile` that defines the files you want to include and compile, and instead of typing a long command or set of commands you simply type `make` and the system will execute a compile based on the definitions in the `makefile`. Here is a tutorial that talks more about this system: `http://www.cs.colby.edu/maxwell/courses/tutorials/maketutor/`, or `http://mrbook.org/tutorials/make/`.

Objective complete – mini debriefing

Now you are equipped to edit and create your own programming files. The next chapters will provide you with lots of opportunity to practice your skills as you translate lines of code into cool robotic capabilities.

Mission accomplished

It is always a bit difficult to try new things. If this is your first attempt at programming, you might feel a bit uncomfortable as I ask you to create or edit files. However, I will try to give you explicit instructions on what to type so that you can be successful. There is one major challenge with working with computers. They always do exactly what you tell them to do, not necessarily what you wanted them to do. So if you encounter problems, check several times to make sure that your code matches the example exactly. Now, on to some actual coding!

A challenge

If you are going to do a significant amount of coding you'll want to install an IDE, or Integrated Development Environment. These environments make it much easier to see, edit, compile, and debug your programs. The most popular of these programs in the Linux world is called Eclipse. If you'd like to know more, start with a Google search, or go to `http://www.eclipse.org/`.

3

Providing Speech Input and Output

Now that your BeagleBone Black is up and operating, you can give your project many different basics of functionality that are really cool. We're going to start with speech; it is a good basic project and offers several examples of adding capability in both HW and SW. So buckle up and get ready to learn the basics of interfacing with your board by facilitating speech.

Mission briefing

You'll be adding a microphone and speaker to our basic board, and you'll add functionality so the robot can recognize voice commands and respond via the speaker. Additionally, you'll be able to issue voice commands and have the robot respond with an action. When you're freed from typing in commands, you can interact with your projects in an impressive way. This project will require adding both HW and SW.

Why is it awesome?

Interfacing with your projects via speech is more fun than typing in commands, and it allows interaction with our project without using a keyboard or mouse. Besides, what self-respecting robot wants to carry around a keyboard? No, you want to interact in natural ways with your projects, and this project will teach you how. Interfacing via speech also helps you find your way around the board, learn the available free functionality, and become familiar with the community of functionality developers.

Your objectives

Your objectives are as follows:

- ▸ Hooking up the HW to make and input sound

- ▸ Using eSpeak to allow your projects to respond in a robot voice

- ▸ Using PocketSphinx to interpret your commands

- ▸ Providing the capability to interpret your commands and have your robot initiate an action

Downloading the example code and colored images

You can download the example code files and colored images for this Packt book you have purchased from your account at http://www.packtpub.com. If you purchased this book elsewhere, you can visit http://www.packtpub.com/support and register to have the files e-mailed directly to you.

Mission checklist

Before beginning this project, you'll need a working BeagleBone Black system that connects to power and the Internet (see *Chapter 1, Getting Started with the BeagleBone Black*, for instructions). Additionally, this project requires a USB microphone/speaker adapter. The board itself does not have either an audio out or audio in. The HDMI output does support audio, but most of your projects will not be connected to video monitors with speaker capability.

You'll need three pieces of HW:

- ▸ A USB device that supports microphone in and speaker out (see the following image)

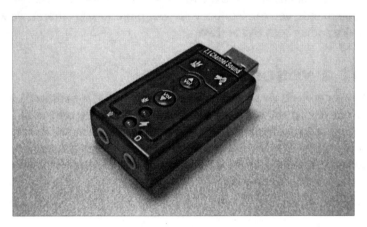

► A microphone that can plug into the USB device (see the following image)

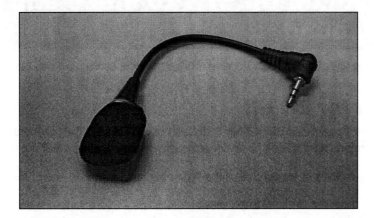

► A powered speaker that can plug in to the USB device (see the following image)

Fortunately, these devices are very inexpensive and widely available. Make sure the speaker is powered because your board will generally not be able to drive a passive speaker with enough power for your applications. A speaker can use either internal battery power or an externally powered USB hub. Many of your projects will require a powered USB hub, so it's a good investment.

Hooking up the HW to make and input sound

For this task, you are going to hook up your HW so that you can record and play sound.

Prepare for lift off

Reassemble your BeagleBone Black. Plug in the LAN cable. Connect the powered USB hub and plug in the microphone/speaker USB device. Also plug in your speakers and the microphone.

Plug in the power. You can execute all of the following instructions in one of the several ways.

If you are still connected to the display, keyboard, and mouse, log in to the board, use `startx` to start Xfce (your windowing system), and then open a terminal window.

If you are only connected via LAN, you can do all of this using an SSH terminal window, so as soon as your board flashes that it has power (look for the heartbeat LED), open up an SSH terminal window using PuTTY or some similar terminal emulator. Once the terminal window comes up, log in with your username and password. Now type in `cat /proc/asound/cards`. You should see the following response:

```
ubuntu@ubuntu-armhf: ~                                        _ □ ✕

login as: ubuntu
ubuntu@157.201.194.187's password:
Welcome to Ubuntu 12.04.2 LTS (GNU/Linux 3.8.13-bone20 armv7l)

 * Documentation:  https://help.ubuntu.com/
Last login: Sat Jan  1 00:03:11 2000 from grimmetr.c.byui.edu
ubuntu@ubuntu-armhf:~$ cat /proc/asound/cards
 0 [Black          ]: TI_BeagleBone_B - TI BeagleBone Black
                      TI BeagleBone Black
 1 [Device         ]: USB-Audio - USB PnP Sound Device
                      USB PnP Sound Device at usb-musb-hdrc.1.auto-1, full speed
ubuntu@ubuntu-armhf:~$ []
```

Notice that the system thinks there are two possible audio devices. The first is the HDMI sound device, and the second is your USB audio plugin. Now you can use the USB card to both create and record sound.

Engage thrusters

First, let's play some music to test that the USB sound device is working. You'll need to configure your system to look for your USB card and use it as the default to play and record sound. To do this, you'll need to add a couple of libraries to your system. The first are some ALSA libraries. ALSA stands for Advanced Linux Sound Architecture, and it is going to enable your sound system on the BeagleBone Black.

First install the two libraries associated with ALSA by typing `sudo apt-get install alsa-base alsa-utils`. Then also install some files that help provide the sound library by typing `sudo apt-get install libasound2-dev`.

If your system already contains these libraries, Linux will simply tell you that they are already installed or that they are up-to-date. After installing both libraries, reboot your BeagleBone Black. It takes time, but the system needs a reboot after new libraries or HW are installed.

Now you'll use an application called alsamixer to control the volume of both the input and the output of your USB sound card. Type `alsamixer` at the prompt. You should see a screen that looks like the following:

Press *F6* and select your USB sound device using the arrow keys. You should see a screen that looks like the following:

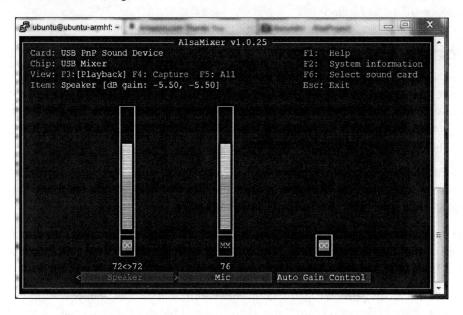

You can use the arrow keys to set the volume for both the speakers and the microphone. Use the *m* key to unmute the microphone. In the preceding screenshot **MM** is mute and ∞ is unmute. Make sure your settings look like the following:

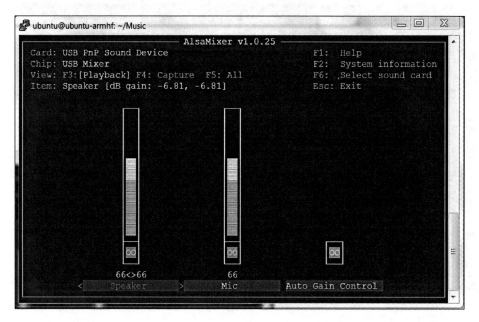

Let's make sure your system knows about your USB sound device. At the prompt, type `aplay -l`. You should see the following:

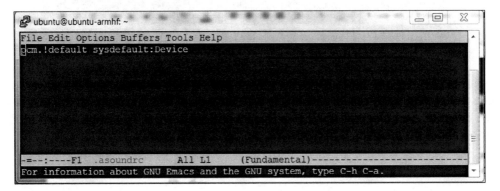

```
ubuntu@ubuntu-armhf: ~
ubuntu@ubuntu-armhf:~$ aplay -l
**** List of PLAYBACK Hardware Devices ****
card 0: Black [TI BeagleBone Black], device 0: HDMI nxp-hdmi-hifi-0 []
  Subdevices: 1/1
  Subdevice #0: subdevice #0
card 1: Device [USB PnP Sound Device], device 0: USB Audio [USB Audio]
  Subdevices: 1/1
  Subdevice #0: subdevice #0
ubuntu@ubuntu-armhf:~$ []
```

If this did not work, try `sudo aplay -l`. Once you have added the libraries, you'll need to add a file. You are going to add a file in your home directory with the name `.asoundrc`. This will be read by your system and used to set your default configuration. To do this:

1. Open the file named `.asoundrc` using your favorite editor.

2. Type in `pcm.!default sysdefault:Device`.

3. Save the file.

The file should look like the following:

```
ubuntu@ubuntu-armhf: ~
File Edit Options Buffers Tools Help
pcm.!default sysdefault:Device




-=--:----F1  .asoundrc       All L1      (Fundamental)-------------------
For information about GNU Emacs and the GNU system, type C-h C-a.
```

This will tell the system to use your USB device as default. Once you have completed this, reboot your system again.

Now, you'll play some music. To do this, you need a sound file and a way to play it. I used WinScp from my Windows machine to transfer a simple `.wav` file to the `Music` subdirectory on my BeagleBone Black. You are going to use an application called aplay to play your sound. Go to the `Music` directory and you should see the music file by simply typing `ll` (list long, this will give you all the details of your files). Mine looks like the following:

```
ubuntu@ubuntu-armhf: ~/Music
ubuntu@ubuntu-armhf:~$ cd ./Music
ubuntu@ubuntu-armhf:~/Music$ ll
total 3496
drwxr-xr-x  2 ubuntu ubuntu    4096 Jan  1 2000 ./
drwxr-xr-x 19 ubuntu ubuntu    4096 Jan  1 2000 ../
-rw-rw-r--  1 ubuntu ubuntu 3568684 Mar 23 2013 Dance.wav
ubuntu@ubuntu-armhf:~/Music$ 
```

Now type `aplay Dance.wav` to see if you can play music using the generic aplay music player. You will see, and hopefully hear, the following result:

```
ubuntu@ubuntu-armhf: ~/Music
ubuntu@ubuntu-armhf:~$ cd ./Music
ubuntu@ubuntu-armhf:~/Music$ ll
total 3496
drwxr-xr-x  2 ubuntu ubuntu    4096 Jan  1 2000 ./
drwxr-xr-x 19 ubuntu ubuntu    4096 Jan  1 2000 ../
-rw-rw-r--  1 ubuntu ubuntu 3568684 Mar 23 2013 Dance.wav
ubuntu@ubuntu-armhf:~/Music$ aplay Dance.wav
Playing WAVE 'Dance.wav' : Signed 16 bit Little Endian, Rate 44100 Hz, Stereo
ubuntu@ubuntu-armhf:~/Music$ 
```

If you aren't hearing any music, check the volume you set with alsamixer and the power to your speaker. Also, aplay can be a bit finicky with the type of files it accepts, so you may have to try different `.wav` files until it will play. One more thing to try, if the system doesn't seem to know about the program, is to type `sudo aplay Dance.wav`.

Now that you can play the sound, let's record some sound. To do this, you're going to use the arecord program. At the prompt, type `arecord -d 5 -r 48000 test.wav`. This will record five seconds of sound at a 48000 Hz sample rate. Once you have typed the command, either speak into the microphone or make some other recognizable sound. You should see the following in the terminal:

```
ubuntu@ubuntu-armhf: ~/Music
ubuntu@ubuntu-armhf:~$ cd ./Music
ubuntu@ubuntu-armhf:~/Music$ ll
total 3496
drwxr-xr-x  2 ubuntu ubuntu    4096 Jan  1  2000 ./
drwxr-xr-x 19 ubuntu ubuntu    4096 Jan  1  2000 ../
-rw-rw-r--  1 ubuntu ubuntu 3568684 Mar 23  2013 Dance.wav
ubuntu@ubuntu-armhf:~/Music$ aplay Dance.wav
Playing WAVE 'Dance.wav' : Signed 16 bit Little Endian, Rate 44100 Hz, Stereo
ubuntu@ubuntu-armhf:~/Music$ arecord -d 5 -r 48000 test.wav
Recording WAVE 'test.wav' : Unsigned 8 bit, Rate 48000 Hz, Mono
ubuntu@ubuntu-armhf:~/Music$ ls
Dance.wav  test.wav
ubuntu@ubuntu-armhf:~/Music$ 
```

Once you create the file, play it with aplay. Type `aplay test.wav`, and you should hear the recording. If you can't hear your recording, check alsamixer to make sure your speakers and microphone are both unmuted.

Objective complete – mini debriefing

Now you can play music or other sound files using your BeagleBone Black. You can change the volume of your speaker and record your voice or other sounds on the system. You're ready for the next step.

Classified intel

Ubuntu offers a number of different music and sound recording program options that are more full-featured than aplay and arecord. If you'd like, spend some time researching them on the Internet and install them onto your system. Several work well with your Xfce installation as well, although you may need to configure your windowing system to know about and use your USB sound device.

Using eSpeak to allow your projects to respond in a robotic voice

Sound is an important tool in your robotic toolkit, but you will want to do more than just play music. Let's allow your robot to speak.

Prepare for lift off

Now that you can both get sound in and out of your BeagleBone Black, let's do something useful with this capability. You're going to start by enabling eSpeak, an open source application that provides you with a computer voice.

Engage thrusters

eSpeak is an open source voice generation application. To get this free functionality, you'll need to do the following.

Download the eSpeak library by typing `sudo apt-get install espeak`. You'll probably have to accept the additional size space that the application requires, but this is fine based on your SD card size. The download may take a while, but the prompt will reappear when it is complete.

Now, let's see if the BeagleBoard Black has a voice. Type the following command: `espeak "hello"`. The speaker should emit a computer voiced "hello". If it is does not, check the speakers and volume level.

Now that you have a computer voice, you need to customize it. eSpeak offers a fairly complete set of customization features, including a large number of languages, "voices", and other options. To access these, you can type in the options at the command line. For example, type in `espeak -v+f3 "hello"`, and you should hear a female voice. Add a Scottish accent by typing `espeak -ven-sc+f3 "hello"`. My personal favorite is the West Midlands accent using a female voice: `espeak -ven-sc+f3 "hello"`. Now that you have your desired voice, you can set it as the default, so you don't always have to include it in the command line.

To set the default, go to the default file definition for eSpeak, which is in the `/usr/share/espeak-data/voices` directory. You should see something like the following:

```
ubuntu@ubuntu-armhf: /usr/share/espeak-data/voices
ubuntu@ubuntu-armhf:/usr/share/espeak-data/voices$ ls
!v   cs        default~  es-la  hi       id   ku  ml  pt-pt  sr    tr
af   cy        el        et     hr       is   la  nl  ro     sv    vi
bg   da        en        fi     hu       it   lv  no  ru     sw    zh
bs   de        eo        fr     hy       ka   mb  pl  sk     ta    zh-yue
ca   default   es        fr-be  hy-west  kn   mk  pt  sq     test
ubuntu@ubuntu-armhf:/usr/share/espeak-data/voices$ []
```

The default file is the one that eSpeak uses to choose a voice. To get your desired voice, en-wm with a female tone, you are going to combine two files into the default file. The first file, the female tone, is in the `!v` directory. Type `\!v` whenever you want to specify this directory. You need to type the `\` character because `!` is a special character in Linux, and if you want to use it as a regular old character, you need to put a `\` before it. Before combining the two files, copy the current default into a file called `default.old`, so that it can be retrieved later if needed. The next step is to copy the `f3` voice into your default file. Type this command: `sudo cp ./\!v/f3 default`. Now edit this file. It should look like the following:

```
ubuntu@ubuntu-armhf: /usr/share/espeak-data/voices
File Edit Options Buffers Tools Help
language variant
name female3
gender female

pitch 140 240
formant 0 105   80 150
formant 1 120   75 150 -50
formant 2 135   70 150 -250
formant 3 125   80 150
formant 4 125   80 150
formant 5 125   80 150
formant 6 120   70 150
formant 7 110   70 150
formant 8 110   70 150

stressAmp 18 18 20 20 20 20 20 20
//breath 0 2 4 4 4 4 4 4
breath 0 2 3 3 3 3 3 2
echo 120 10
roughness 4

-=--:----F1  default        Top L1      (Fundamental)----------------------
For information about GNU Emacs and the GNU system, type C-h C-a.
```

This has all the settings for your female voice. The setting for the accent will be in the en-wm file, located in the en directory. Combining the two will give us the following file:

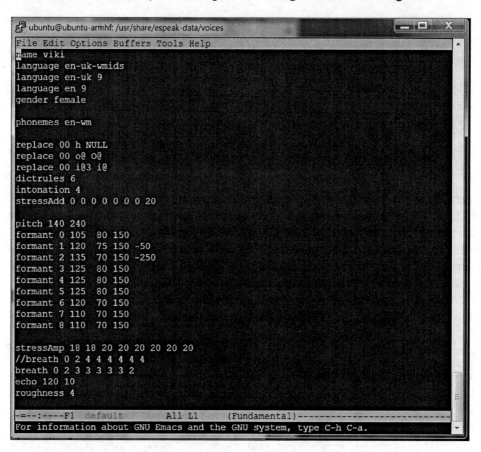

Now you can simply type espeak and get your desired computer voice.

Objective complete – mini debriefing

Now your project can speak. Simply type espeak followed by the text you want to speak in quotes, and out comes your speech. If you want to read an entire text file, you can do that as well, using the -f option and then typing the name of the file. Try this by using your editor to create a text file called speak, then typing this command: espeak -f speak.txt.

Classified intel

There are lots of choices with respect to eSpeak. Feel free to play around and choose your favorite. Then edit the default file to set it to that voice. Don't expect, however, that you'll get the kind of voices that you hear from computers in the movies. Those are actors, not computers, although one day we will hopefully get to the point where computers will sound a lot like people.

Using PocketSphinx to interpret your commands

Sound is cool, and speech is even cooler, but you also want to be able to communicate with your projects through voice commands. This section will show you how to add speech recognition to your robotic projects.

Prepare for lift off

Now that your project can speak, you want it to listen as well. This isn't nearly as simple as the speaking part, but thankfully you have some significant help from the development community. You are going to download a set of capabilities called **PocketSphinx**, which will allow your project to listen to your commands.

Engage thrusters

The first step is downloading the PocketSphinx capability. Unfortunately, this is not quite as user friendly as the eSpeak process, so follow along carefully.

Go to the Sphinx website, hosted by Carnegie Mellon University at `http://cmusphinx.sourceforge.net/`. This is an open source project that provides you with the speech recognition SW. With your smaller embedded system, you will be using the PocketSphinx version of this code.

You will have to download two pieces of SW modules: sphinxbase and PocketSphinx. Select the download option at the top of the page, and then find the latest version of both of these packages. Download the `.tar.gz` version of these and move them to the `/home/ubuntu` directory of your BeagleBone Black. However, before you build these, you need two libraries.

The first library is `libasound2-dev`. If you skipped the first two objectives of this project, you'll need to download it now using `sudo apt-get install libasound2-dev`. If you're unsure that it's installed, try it again. The system will warn you if it's already installed.

The second is a library called `Bison`. This is a general purpose, open source parser that will be used by PocketSphinx. To get this package, type `sudo apt-get install bison`.

Once everything is installed and downloaded, you can build PocketSphinx. First, your home directory should look like the following, with the `tar.gz` files of both `pocketsphinx` and `sphinxbase`:

```
ubuntu@ubuntu-armhf: ~
ubuntu@ubuntu-armhf:~$ ls
Desktop     Music     Templates                    speak
Documents   Pictures  Videos                       sphinxbase-0.8.tar.gz
Downloads   Public    pocketsphinx-0.8.tar.gz
ubuntu@ubuntu-armhf:~$
```

To unpack and build the sphinxbase, type `sudo tar -xzvf sphinx-base-0.x.tar.gz`, where `x` is the version number; in my case it is `8`. This should unpack all the files from the archive into a directory called `sphinxbase-0.x`. Now change directory to `sphinxbase-0.x`. Listing the files should show something like the following:

```
ubuntu@ubuntu-armhf: ~/sphinxbase-0.8
ubuntu@ubuntu-armhf:~/sphinxbase-0.8$ ls
AUTHORS      NEWS          config.sub    include       sphinxbase.pc.in
COPYING      README        configure     install-sh    sphinxbase.sln
ChangeLog    aclocal.m4    configure.in  ltmain.sh     src
INSTALL      autogen.sh    depcomp       m4            test
Makefile.am  config.guess  doc           missing       win32
Makefile.in  config.rpath  group         python        ylwrap
ubuntu@ubuntu-armhf:~/sphinxbase-0.8$
```

To build the application, start by issuing the command: `./configure --enable-fixed`. This command will check to make sure everything is fine with the system and then configure a build. When I first attempted this command, I got the following error:

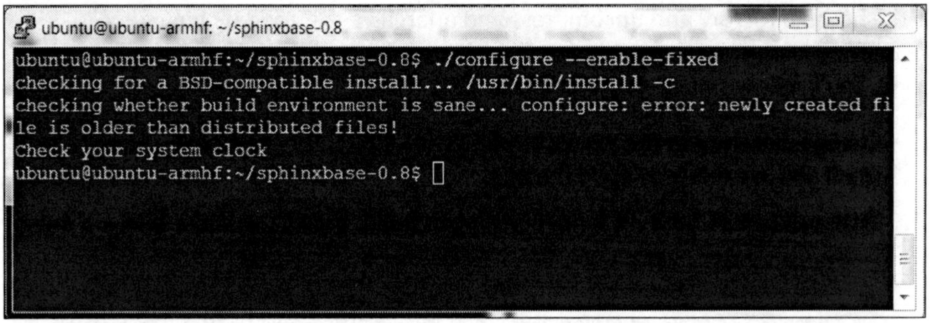

This highlighted an interesting problem. The time and date on my BeagleBone Black was not set to the current time and date. The BeagleBone Black does not have a battery like your PC, so it cannot store a date. Issuing the date command confirmed this as shown in the following screenshot:

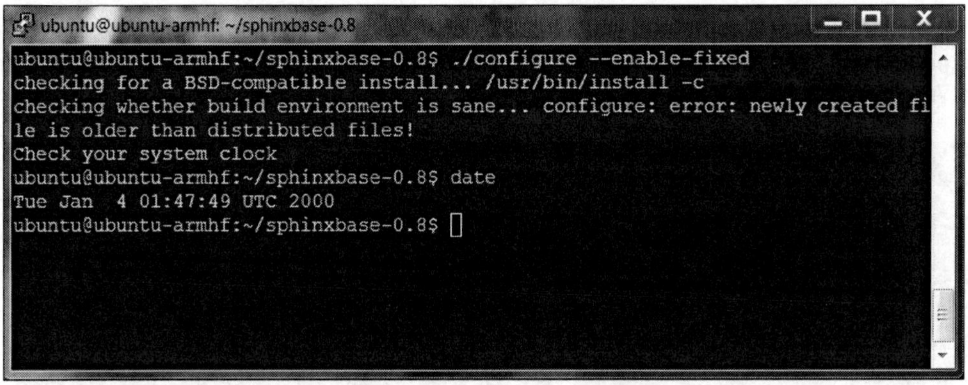

If you need to set the current date and time, do that by issuing the command `sudo date nnddhhmmyyyy.ss`, where `nn` is the month, `dd` is the day, `hh` is the hour, `mm` is the minute, `yyyy` is the year, and `ss` is the seconds. This will set the date to the desired date. Now you can reissue the `./configure --enable-fixed` command.

One final install will enable you to use makefiles to compile your code. This library is `build-essential`. Install this by typing `sudo apt-get install build-essential`. Now you are ready to actually build the sphinxbase codebase. This is a two-step process:

1. Type `make`, and the system will build all the executable files.

2. Type `sudo make install`, and it will install all the executables onto the system.

Now you need to make the second part of the system: the PocketSphinx code itself.

Go to the home directory and uncompress and unarchive the code by typing `tar -xzvf pocketsphinx-0.8.tar.gz`. The files should now be unarchived, and you can now build the code. You'll follow similar steps for these files:

1. Change directory to the `pocketsphinx-0.8` directory and type `./configure` to see if you are ready to build the files.

2. Type `make`, wait for a bit for everything to build, then type `sudo make install`.

> Several possible additions to your library installation will be useful later if you are going to use your PocketSphinx capability with Python as a coding language. You can install python-dev using `sudo apt-get install python-dev` and Cython using `sudo apt-get install cython`. You can also choose to install pkg-config, a utility that can sometimes help do with complex compilations. Install it using `sudo apt-get install pkg-config`.

Once the installation is complete, you'll need to let the system know where your files are. To do this you will need to edit the `/etc/ld.so.conf` as root. Add one line to the end of the file, /usr/local/lib, so that your file looks like this:

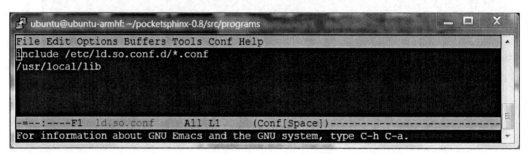

Now type `/sbin/ldconfig`, and the system will now be aware of your PocketSphinx libraries.

Everything is installed, so you can try your speech recognition. Change directory to the `bring this all on one line` directory to try a demo program. Type `pocketsphinx_continuous`. This program takes in input from the microphone and turns it into speech. After running the command, you'll get a lot of irrelevant information and then see the following:

```
INFO: ngram_model_dmp.c(288):    436879 = LM.bigrams(+trailer) read
INFO: ngram_model_dmp.c(314):    418286 = LM.trigrams read
INFO: ngram_model_dmp.c(339):     37293 = LM.prob2 entries read
INFO: ngram_model_dmp.c(359):     14370 = LM.bo_wt2 entries read
INFO: ngram_model_dmp.c(379):     36094 = LM.prob3 entries read
INFO: ngram_model_dmp.c(407):       854 = LM.tseg_base entries read
INFO: ngram_model_dmp.c(463):      5001 = ascii word strings read
INFO: ngram_search_fwdtree.c(99): 788 unique initial diphones
INFO: ngram_search_fwdtree.c(147): 0 root, 0 non-root channels, 60 single-phone
words
INFO: ngram_search_fwdtree.c(186): Creating search tree
INFO: ngram_search_fwdtree.c(191): before: 0 root, 0 non-root channels, 60 singl
e-phone words
INFO: ngram_search_fwdtree.c(326): after: max nonroot chan increased to 13428
INFO: ngram_search_fwdtree.c(338): after: 457 root, 13300 non-root channels, 26
single-phone words
INFO: ngram_search_fwdflat.c(156): fwdflat: min_ef_width = 4, max_sf_win = 25
INFO: continuous.c(371): pocketsphinx_continuous COMPILED ON: Jul 18 2013, AT: 0
9:06:25

Warning: Could not find Mic element
Warning: Could not find Capture element
READY....
```

Even though it tells you that it can't find your microphone element or a capture element, if you have set things up as previously described, you should be ready to give it a command. Say "hello" into the microphone. When it senses that you have stopped speaking, it will process your speech, give lots of irrelevant information again, but should eventually show this screen:

```
INFO: ngram_search_fwdtree.c(1557):     5469 words for which last channels evalu
ated (89/fr)
INFO: ngram_search_fwdtree.c(1560):     41837 candidate words for entering last p
hone (685/fr)
INFO: ngram_search_fwdtree.c(1562): fwdtree 1.97 CPU 3.224 xRT
INFO: ngram_search_fwdtree.c(1565): fwdtree 3.41 wall 5.591 xRT
INFO: ngram_search_fwdflat.c(302): Utterance vocabulary contains 145 words
INFO: ngram_search_fwdflat.c(937):     2317 words recognized (38/fr)
INFO: ngram_search_fwdflat.c(939):     96101 senones evaluated (1575/fr)
INFO: ngram_search_fwdflat.c(941):    182559 channels searched (2992/fr)
INFO: ngram_search_fwdflat.c(943):      8014 words searched (131/fr)
INFO: ngram_search_fwdflat.c(945):      6797 word transitions (111/fr)
INFO: ngram_search_fwdflat.c(948): fwdflat 0.69 CPU 1.124 xRT
INFO: ngram_search_fwdflat.c(951): fwdflat 0.69 wall 1.132 xRT
INFO: ngram_search.c(1266): lattice start node <s>.0 end node </s>.52
INFO: ngram_search.c(1294): Eliminated 0 nodes before end node
INFO: ngram_search.c(1399): Lattice has 281 nodes, 2180 links
INFO: ps_lattice.c(1365): Normalizer P(O) = alpha(</s>:52:59) = -478171
INFO: ps_lattice.c(1403): Joint P(O,S) = -484043 P(S|O) = -5872
INFO: ngram_search.c(888): bestpath 0.06 CPU 0.107 xRT
INFO: ngram_search.c(891): bestpath 0.07 wall 0.114 xRT
000000001: hello
READY....
```

Notice the **000000001: hello**. It recognized your speech! You can try other words and phrases too. The system is very sensitive, so it may pick up background noise. You are also going to find that it is not very accurate. If you'd like to improve the accuracy, see the *Classified intel* section. To stop the program, type cntrl-c.

Objective complete – mini debriefing

Your system can understand your speech input! In the next section of this project, you'll learn how to use this input to have the project respond.

Classified intel

There are two ways to make the system more accurate. One is to train the system to more accurately understand your voice. It is a bit complex, and if you want to know more go to Carnegie Mellon University's (CMU) PocketSphinx website.

The second way to improve accuracy is to limit the number of words that your system uses to determine what you are saying. The default has literally thousands of word possibilities, so if two words are close, it may choose the wrong word. To avoid this, you can make your own grammar to restrict the words it has to choose from.

The first step is to create a file with the words or phrases you want the system to recognize. Then you use a web tool to create two files that the system will use to define your grammar. I'll do this through the vncserver, since I'll need to use a web browser. The next step is to create a file called `grammar.txt` and insert the text shown in the following screenshot:

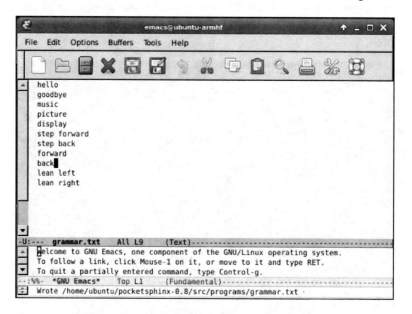

Now you must use the CMU web tool to turn this file into two files that the system can use to define its dictionary. On my system, I have already installed Firefox using `sudo apt-get install firefox`. So, now I can open a web browser window and go to this URL: `http://www.speech.cs.cmu.edu/tools/lmtool-new.html`. If I hit the **browse** button, I can then find and select the file. It should look something like the following:

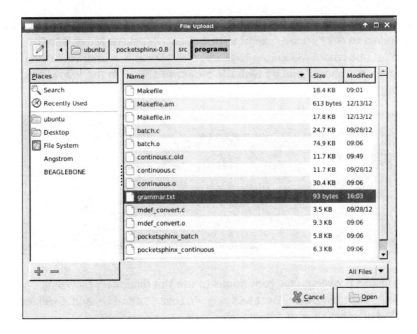

Open the `grammer.txt` file, then on the web page select **COMPILE KNOWLEDGE BASE**, and the following window should pop up:

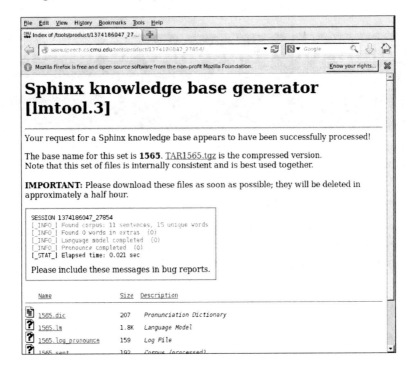

You need to download the .tgz file the tool created, in this case the TAR1565.tgz file. This will download to your /home/ubuntu/Download directory. Move it to the /home/ubuntu/pocketsphinx-0.8/src/programs directory and unarchive it using tar -xzvf and the filename. You should end up with the programs shown in the following screenshot in the directory:

Now you can invoke pocketsphinx_continuous to use this dictionary by typing pocketsphinx_continuous -lm 1565.lm -dict 1565.dic, and it will look in that directory to find matches to your commands.

You can also do this on your remote computer using Windows, creating the file in a text editor such as WordPad. Once you have created the required grammar files, you can download them to your BeagleBone Black using WinScp.

Providing the capability to interpret your commands and have your robot initiate an action

Now that your robot can both speak and listen, let's see if you can make it respond to your commands.

Prepare for lift off

Now that the system can both hear and speak, you want to provide the capability to respond to your speech and execute some commands based on the speech input. Now you're going to configure the system to respond to simple commands.

Engage thrusters

In order to respond, you're going to edit the `continuous.c` code in the `/home/ubuntu/src/programs` directory. You could create your own C file, but this file is already set up in the makefile system and is an excellent starting spot. I like to make a copy of the current file into `continuous.c.old`, so I can always get back to the starting program if it is required. Then you will need to edit the `continuous.c` file. It is very long and a bit complicated, but you are specifically looking for the following section in the code:

```
File Edit Options Buffers Tools C Help
        while (ad_read(ad, adbuf, 4096) >= 0);
        cont_ad_reset(cont);

        printf("Stopped listening, please wait...\n");
        fflush(stdout);
        /* Finish decoding, obtain and print result */
        ps_end_utt(ps);
        hyp = ps_get_hyp(ps, NULL, &uttid);
        printf("%s: %s\n", uttid, hyp);
        fflush(stdout);

        /* Exit if the first word spoken was GOODBYE */
        if (hyp) {
            sscanf(hyp, "%s", word);
            if (strcmp(word, "goodbye") == 0)
                break;
        }

        /* Resume A/D recording for next utterance */
        if (ad_start_rec(ad) < 0)
            E_FATAL("Failed to start recording\n");
    }

    cont_ad_close(cont);
-=--:----F1  continuous.c    80% L327    (C/l Abbrev)---------------
```

In this section of the code, the word has already been decoded and is held in the variable `hyp`. You can add code here to make your system do things based on the value associated with the word you have decoded. First, let's try adding the capability to respond to "hello" and "goodbye" to see if you can get the program to stop. Make the following changes to the code:

1. Find the comment `/* Exit`, if the first word spoken was `GOODBYE */`.

2. In the statement `if (strcmp(word, "good bye") == 0)`, change the `good bye` to `GOODBYE`.

3. Put `{ }` around the `break;` statement and add the following statement just before the `break;`: `system ("espeak" \"good bye\"");`.

4. Add another else if statement to the clause by typing `else if (strcmp(hyp, "HELLO") == 0)`. Add `{ }` after the else if statement and inside the brackets, `system ("espeak" \"good bye\"");`.

The file should now look like this:

```
File Edit Options Buffers Tools C Help
        fflush(stdout);
        /* Finish decoding, obtain and print result */
        ps_end_utt(ps);
        hyp = ps_get_hyp(ps, NULL, &uttid);
        printf("%s: %s\n", uttid, hyp);
        fflush(stdout);
        /* Exit if the first word spoken was GOODBYE */
        if (hyp) {
            sscanf(hyp, "%s", word);
            if (strcmp(hyp, "GOODBYE") == 0)
                {
                system("espeak \"good bye\"");
                break;
                }
            else if (strcmp(hyp, "HELLO") == 0)
                {
                system("espeak \"hello\"");
                }
        }

        /* Resume A/D recording for next utterance */
        if (ad_start_rec(ad) < 0)
            E_FATAL("Failed to start recording\n");
    }
-=--:----F1   continuous.c   79% L319   (C/1 Abbrev)-----------------------
```

Now you need to rebuild your code. Since the make system already knows how to build the program pocketsphinx_continuous, anytime you make a change to the continuous.c file, it will rebuild the application. Simply type make, and the file will compile and create a new version of pocketsphinx_continuous. To run your new version, type ./pocketsphinx_continuous. Make sure you type the ./ at the start; if not, the system has another version of pocketsphinx_continuous in the library and will run that.

If everything is set correctly, saying "hello" should result in a response of "hello" from your BeagleBone Black. Saying "good bye" should elicit a response of "good bye," as well as shutting down the program. Notice the system command can be used to run any program that runs with a command line. Now you can use this program to start and run other programs based on the commands.

Objective complete – mini debriefing

Finally, your BeagleBone Black will both listen and respond, and it will execute a command. You are now ready to move on to providing sight for your system.

Classified intel

I'm using the C files that came with PocketSphinx to interact with the system. A set of Python files also came with the system. If you prefer to work in Python, go ahead and explore those examples in the `/home/ubuntu/pocketsphinx-0.8/python` directory.

Mission accomplished

Now your project can both hear and speak. You can use this later when you want to interface with your project without typing commands or using a display. You should also feel more comfortable installing new hardware and software onto your system. You'll be using that skill throughout the book as you look at more complex projects.

A challenge

You'll use this capability to allow your robot to respond to your commands in several of your projects. I've used one example with the system C command. You should try others. You could also start programs and keep them going, interfacing with them without invoking them over and over again at the command line. If you are familiar with Linux, think about the messaging protocols that you could use to interface two running programs.

4

Allowing the BeagleBone Black to See

Your projects can communicate via voice; now you are going to add vision with a webcam. You'll use this functionality in lots of different applications. Fortunately, adding hardware and software for vision is both easy and inexpensive.

Mission briefing

In this chapter, you'll add a USB webcam to your system. Having the standard USB interface on your board opens a wide range of amazing possibilities. On top of that, there are several amazing open-source libraries. These offer complex capabilities that you can use in your projects without spending months coding them.

Why is it awesome?

Vision will open a set of possibilities for your project. These can range from simple motion detection to advanced capabilities such as facial recognition, object identification, and even object tracking. The robot can also use vision to detect its surroundings and avoid obstacles.

Your objectives

In this chapter we will cover:

- ▸ Connecting your USB camera to your BeagleBone Black and viewing the images
- ▸ Downloading and installing OpenCV, a full-featured vision library
- ▸ Using the vision library to detect colored objects

Downloading the example code and colored images

You can download the example code files and colored images for this Packt book you have purchased from your account at http://www.packtpub.com. If you purchased this book elsewhere, you can visit http://www.packtpub.com/support and register to have the files e-mailed directly to you.

Mission checklist

To complete this mission, you'll need a BeagleBone Black with a LAN connection and a 5V power supply. You'll need to add a USB webcam. Try to find a recently manufactured one. You may have an older webcam sitting on your project shelf, but it will probably cause problems, and the money you save will not be worth the frustration. I would stick with webcams from the major players such as Logitech or Creative Labs.

In most cases, you won't need to connect this device through a powered-USB hub; however, if you encounter problems, for example if the system does not recognize that your webcam is connected, realize that the lack of USB power could be the problem.

Connecting the USB camera to the BeagleBone Black and viewing the images

Our first step in enabling computer vision is connecting the USB camera to the USB port. I have a new Logitech HD 720 camera as my example.

Prepare for lift off

To access the USB webcam, I like to use a program called **guvcview**. Install this by typing `sudo apt-get install guvcview`.

Engage thrusters

Connect your USB camera and make sure your LAN cable is plugged in. Then apply power to the BeagleBone Black. After the system is booted, you can check to see if the BeagleBone Black has found your USB camera. Go to the /dev directory and type ls. You should see the output as shown in the following screenshot:

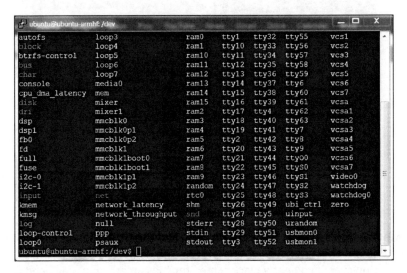

Look for the **video0** device, the webcam. If you see it, the system knows your camera is there.

Now let's use guvcview to see the output of the camera. Since this will need to output some graphics, you either need to use a monitor connected to the board as well as a keyboard and a mouse, or you can use vncserver. If you are going to use vncserver, make sure you start the server on the BeagleBone Black by typing vncserver via SSH. Then start up your vncviewer as described in *Chapter 1, Getting Started with the BeagleBone Black*. Open a terminal window and type sudo guvcview.

You should see the output as shown in the following screenshot:

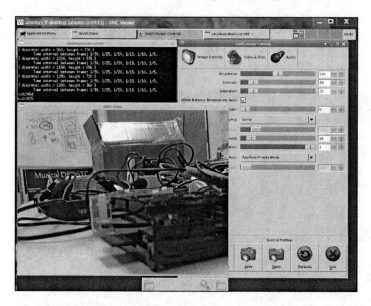

The video window displays what the webcam sees, and the **GUVCViewer Controls** window controls the different characteristics of the camera. The default settings of the Logitech HD 720 work fine. However, if you get a black screen for the camera, you may need to adjust the settings. Click on the **GUVCViewer Controls** window and the **Video & Files** tab. You will see a window where you can adjust the settings for your camera.

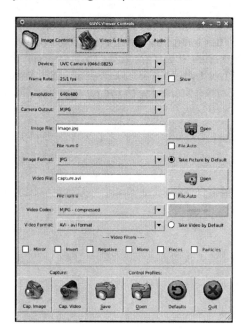

The most important setting is the **Resolution**. If you see a black screen, adjust the resolution down; often this will resolve the issue. This window will also tell you what resolutions are supported for your camera. Also, you can display the frame rate by checking the box at the right of the **Frame Rate** setting. Be aware, however, that if you are going through vncviewer, the refresh rate will be much slower than if you're using the BeagleBone Black and a monitor directly.

Once you have the camera up and running and a desired resolution set, you can go on to download and install OpenCV.

Objective complete – mini debriefing

Your system can now "see" the outside world. Guvcview can actually capture images or video and store them as files, but OpenCV provides a full-featured set of image processing capabilities as well.

Classified intel

You can connect more than one webcam to the system. Follow the same steps, but connect to cameras via a USB hub. List the devices in the /dev directory. Use guvcview to see the different images. One challenge, however, is that connecting too many cameras can overwhelm the bandwidth of the USB port.

Downloading and installing OpenCV – a full-featured vision library

Now that you have your camera connected, you can begin to access some amazing capabilities that have been provided by the open source community. The most popular of these for computer vision is OpenCV.

Prepare for lift off

Now you need to install OpenCV, a complete vision library that provides tools for you to use to capture, process, and save your images. Before you do this, you need to expand the partition on your SD card so that you can download all the applications that you need. When you wrote the Linux operating system to your SD card, you copied a 2 GB image; so now your card thinks it is only a 2 GB card, no matter what size it really is. You need to reclaim this space.

To do this, you'll need to issue some fairly cryptic commands, but you can use the defaults, so it will be straightforward. First, open a terminal window. The card I am using is an 8 GB card, so if your card is of a different size, you might not see the exact same numbers. Fortunately you'll be using default values throughout the process, so you won't need to know anything special about your card. To begin, type `sudo su`, and then enter your password. Then follow these steps:

1. Type `ll /dev/mmcblk*` and you should see output similar to the following screenshot:

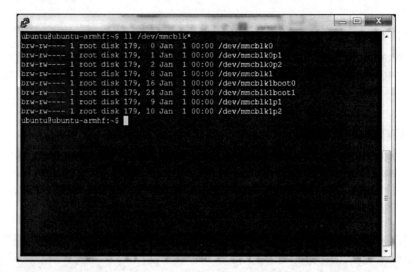

2. Now you are going to make changes to the `mmcblk0` device. Type `fdisk /dev/mmcblk0`.

3. Enter the `p` command and you should see the details of the different types of storage that you currently have available, similar to the output shown in the following screenshot:

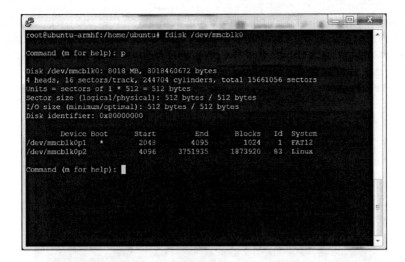

4. You are going to expand the second device, /dev/mmcblk0p2. To do this you are going to delete the partition, then create a new partition. The information that exists on your SD card should be preserved throughout this process, however. Enter d at the prompt, then 2, for partition 2. Now enter p again and the following screenshot will be displayed:

```
Sector size (logical/physical): 512 bytes / 512 bytes
I/O size (minimum/optimal): 512 bytes / 512 bytes
Disk identifier: 0x80000000

      Device Boot      Start         End      Blocks   Id  System
/dev/mmcblk0p1    *       2048        4095        1024    1  FAT12
/dev/mmcblk0p2            4096     3751935     1873920   83  Linux

Command (m for help): d
Partition number (1-4): 2

Command (m for help): p

Disk /dev/mmcblk0: 8018 MB, 8018460672 bytes
4 heads, 16 sectors/track, 244704 cylinders, total 15661056 sectors
Units = sectors of 1 * 512 = 512 bytes
Sector size (logical/physical): 512 bytes / 512 bytes
I/O size (minimum/optimal): 512 bytes / 512 bytes
Disk identifier: 0x80000000

      Device Boot      Start         End      Blocks   Id  System
/dev/mmcblk0p1    *       2048        4095        1024    1  FAT12

Command (m for help):
```

5. Now you will create a new partition using defaults so that the partition takes up the entire card. At the Command prompt type an n, then p, then 2 and then hit *Enter* through each choice that the programs request. Your device should now reappear, similar to the following screenshot, with a different size based on the size of your SD card:

```
Partition type:
   p   primary (1 primary, 0 extended, 3 free)
   e   extended
Select (default p): p
Partition number (1-4, default 2): 2
First sector (4096-15661055, default 4096):
Using default value 4096
Last sector, +sectors or +size{K,M,G} (4096-15661055, default 15661055):
Using default value 15661055

Command (m for help): p

Disk /dev/mmcblk0: 8018 MB, 8018460672 bytes
4 heads, 16 sectors/track, 244704 cylinders, total 15661056 sectors
Units = sectors of 1 * 512 = 512 bytes
Sector size (logical/physical): 512 bytes / 512 bytes
I/O size (minimum/optimal): 512 bytes / 512 bytes
Disk identifier: 0x80000000

      Device Boot      Start         End      Blocks   Id  System
/dev/mmcblk0p1   *      2048        4095        1024    1  FAT12
/dev/mmcblk0p2          4096    15661055     7828480   83  Linux

Command (m for help):
```

6. Notice that the second partition is now much larger than the original. Now type w to commit your changes. Now you need to reboot, so type reboot.

7. The final steps will expand the file system. After the system reboots, type sudo su and enter your password. Now type df, the command to see how much disk you have free. You should be able to see the output as shown in the following screenshot:

```
root@ubuntu-armhf:/home/ubuntu# df
Filesystem     1K-blocks    Used Available Use% Mounted on
/dev/mmcblk0p2  1811704  375204   1342804  22% /
devtmpfs         253768       4    253764   1% /dev
none              50784     276     50508   1% /run
none               5120       0      5120   0% /run/lock
none             253912       0    253912   0% /run/shm
/dev/mmcblk0p1     1004     472       532  48% /boot/uboot
root@ubuntu-armhf:/home/ubuntu#
```

8. It is the `/dev/mmcblk0p2` device that you want to resize. Type `resize2fs /dev/mmcblk0p2` and then enter `df`; you should be able to see the output as shown in the following screenshot:

Now your device is ready to use.

Engage thrusters

First, you'll need to download a set of libraries and OpenCV itself. There are several possible steps. I'm going to suggest the ones that I follow to install it on my systems. Once you have booted the system and opened a terminal window, type the following commands in the given order:

1. `sudo apt-get install update`: If you haven't done this in a while, it is a good idea to do this now before you start. You will be downloading a number of new software packages, so it is good to make sure everything is up to date.

2. `sudo apt-get install build-essential`: You have done this in the previous chapter, but if you skipped that part, you are going to need this package.

3. `sudo apt-get install libavformat-dev`: This library provides a way to code and decode audio and video streams.

4. `sudo apt-get install ffmpeg`: This library provides a way to transcode audio and video streams.

5. `sudo apt-get install libcv2.3 libcvaux2.3 libhighgui2.3`: These are the basic OpenCV libraries. Note the number. This will almost certainly change as new versions of OpenCV become available. If Version 2.3 does not work, either try Version 2.4 or google for the latest version of OpenCV.

6. `sudo apt-get install python-opencv`: This is the Python development kit for OpenCV, needed as you are going to use Python.

7. `sudo apt-get install opencv-doc`: This library provides the documentation for OpenCV, just in case you need it.

8. `sudo apt-get install libcv-dev`: This library provides the header files and static libraries to compile OpenCV.

9. `sudo apt-get install libcvaux-dev`: This library provides more development tools for compiling OpenCV.

10. `sudo apt-get install libhighgui-dev`: This is another package that provides header files and static libraries to compile OpenCV.

11. Make sure you are in your home directory, and then type `cp-r/usr/share/doc/opencv-doc/examples`: This will copy all the examples to your home directory.

12. Go to the `examples/c` directory by typing `cd ./examples/c` and type `sh build_all.sh` and you will have a set of executable files that will allow you to see if the system is working.

Now you are ready to try out the OpenCV library. I prefer to use Python when programming simple tasks, so I'll show the Python examples. If you prefer the C examples, feel free to explore. In order to use the Python examples, you'll need one more library. So type `sudo apt-get install python-numpy` as you will need this to manipulate the matrices that OpenCV uses to hold the images.

Now that you have those, you can try one of the Python examples. Change directory to the Python examples by typing `cd /home/ubuntu/examples/python`. In this directory you will find a number of useful examples, we'll only look at the most basic. It is called `camera.py`. You can try running this example; however, to do this you'll either need to have a display connected to the BeagleBone Black, or you can do this over the vncserver connection. Bring up a terminal window and type `python camera.py`. You should see output similar to the following screenshot:

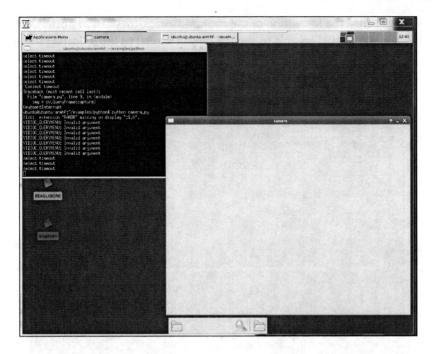

In my case the camera window eventually turned black, but did not show the output from the camera. I found that I needed to change the resolution of the image to the one supported by the camera and OpenCV. To do this you edit the `camera.py` file, and add two lines as shown in the following screenshot:

```
File Edit Options Buffers Tools Python Help
import cv2.cv as cv
import time

cv.NamedWindow("camera", 1)

capture = cv.CaptureFromCAM(0)
cv.SetCaptureProperty(capture, 3, 360)
cv.SetCaptureProperty(capture, 4, 240)

while True:
    img = cv.QueryFrame(capture)
    cv.ShowImage("camera", img)
    if cv.WaitKey(10) == 27:
        break



-=--:-==-F1  camera.py      All L1      (Python)-------------------
For information about GNU Emacs and the GNU system, type C-h C-a.
```

These two lines change the resolution of the captured image to 360 x 240 pixels. Now run `camera.py`, and you should be able to see the output similar to the following screenshot:

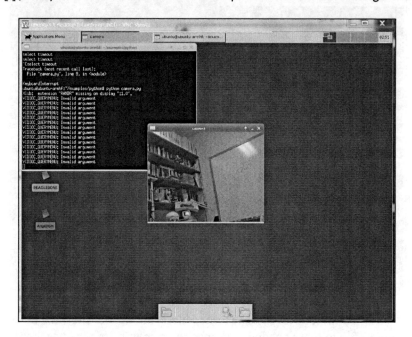

Objective complete – mini debriefing

Your project can now see! You will use this capability to do a number of impressive tasks that will use this vision capability.

Classified intel

You may want to play with the resolution to find the optimum for your application. Bigger images are great—they give you more detailed view of the world—but they also take up significantly more processing power. You'll play with this more as you actually ask your system to do some real image processing. Be careful if you are going to use vnc to understand your system performance as this will significantly slow down the update rate. An image that is twice the size (width/height) will involve four times more processing.

Using the vision library to detect colored objects

Now that you have access to the OpenCV library, let's see what it can do.

Prepare for lift off

OpenCV and your webcam can track objects. This might be useful if you are building a system that needs to track and follow a colored ball. OpenCV makes this amazingly simple by providing some high-level libraries that can help you with this task. I'm going to do this in Python, as I find it much easier to work with than C. If you feel more comfortable in C, these instructions should be fairly easy to translate. Also, performance will be better if implemented in C, so you might create the initial capability in Python, and then finalize the code in C.

Engage thrusters

If you'd like, create a directory to hold your image-based work. From your home directory, create a directory named `imageplay` by typing `mkdir imageplay`. Then change directory to `imageplay` by typing `cd imageplay`.

Once there, let's bring over your `camera.py` file as a starting point by typing `cp /home/ubuntu/examples/python/camera.py ./camera.py`. Now you are going to edit the file until it looks similar to the following screenshot:

Let's look specifically at the changes you need to make to `camera.py`. The first four lines you add are as follows:

```
#Smooth image, then convert the Hue
cv.Smooth(img,img,cv.CV_BLUR,3)
hue_img = cv.CreateImage(cv.GetSize(img), 8, 3)
cv.CvtColor(img,hue_img, cv.CV_BGR2HSV)
```

We are going to use the OpenCV library to first smooth the image, taking out any large deviations. The next two lines create a new image that stores the image in values of **HSV** (**Hue** (color), **Saturation**, and **Value**) instead of the **RGB** (**Red**, **Green**, and **Blue**) pixel values of the original image. Converting to HSV focuses your processing more on the color, as opposed to the amount of light hitting it.

Then we add the following lines of code:

```
#Remove all the pixels that don't match
threshold_img = cv.CreateImage(cv.GetSize(hue_img), 8, 1)
cv.InRangeS(hue_img, (38,120, 60), (75, 255, 255),
            threshold_img)
```

You are going to create yet one more image, this time a black-and-white binary image that is black for any pixel which is not between two certain color values. The (38, 120, 60) and (75, 255, 255) parameters determine the color range. In this case I have a green ball, and I want to detect the color green.

Now run the program. You'll need to either have a display, keyboard, and mouse connected to the board or you can run it remotely using vnc. Run the program by typing sudo python camera.py. You should see a single black image, but move this window and you will expose the original image window as well. Now take your target (I used my green ball) and move it into the frame. You should see output similar to the following screenshot:

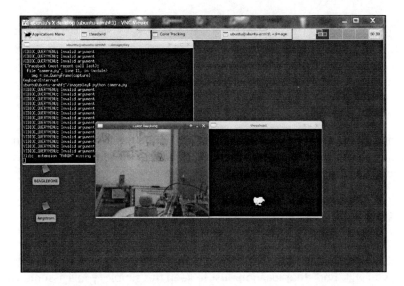

Notice the white pixels in your threshold image showing where the ball is located. You can add more OpenCV code that gives the actual location of the ball. You can actually draw a rectangle around the ball as an indicator in your original image file of the location of the ball. Edit the camera.py file to look like the following screenshot:

```
cv.CvtColor(img,hue_img, cv.CV_BGR2HSV)

    #Remove all the pixels that don't match
    threshold_img = cv.CreateImage(cv.GetSize(hue_img), 8, 1)
    cv.InRangeS(hue_img, (38,160, 60), (75, 256, 256), threshold_img)

    # Find all the areas of color out there
    storage = cv.CreateMemStorage(0)
    contour = cv.FindContours(threshold_img, storage, cv.CV_RETR_CCOMP, cv.CV_C\
HAIN_APPROX_SIMPLE)

    #Step through all the areas
    points = []
    while contour:
        # Get the info about this area
        rect = cv.BoundingRect(list(contour))
        contour = contour.h_next()
        #Check to make sure the area is big enough to be of concern
        size = (rect[2] * rect[3])
        if size > 100:
            pt1 = (rect[0], rect[1])
            pt2 = (rect[0] + rect[2], rect[1] + rect[3])
            #Add a rectangle to the initial image
            cv.Rectangle(img, pt1, pt2, (160, 160, 160))

    cv.ShowImage("Color Tracking", img)
#    cv.ShowImage("threshold", threshold_img)
    if cv.WaitKey(10) == 27:
        break
```

First, add these lines:

```
# Find all the areas of color out there
    storage = cv.CreateMemStorage(0)
    contour = cv.FindContours(threshold_img, storage, cv.CV_RETR_
CCOMP, cv.CV_C\
HAIN_APPROX_SIMPLE)
```

These lines find all the areas on your image that are within the threshold. There may be more than one; so you want to capture them all. Now you will add a `while` loop that will let you step through all the possible contours:

```
#Step through all the areas
points = []
while contour:
```

By the way, it is important to note that if there is another larger green blob in the background, you will "find" that location. Just to keep this simple, you'll assume your green ball to be unique. The next few lines will then get the information for each of your contours. Now, you want to identify the corners. Then you can check to see if the area is big enough to be of concern. If it is, you will add a rectangle to your original image identifying where you think it is:

```
# Get the info about this area
rect = cv.BoundingRect(list(contour))
contour = contour.h_next()
#Check to make sure the area is big enough to be of
    concern
size = (rect[2] * rect[3])
if size > 100:
    pt1 = (rect[0], rect[1])
    pt2 = (rect[0] + rect[2], rect[1] + rect[3])
    #Add a rectangle to the initial image
    cv.Rectangle(img, pt1, pt2, (38, 160, 60))
```

Now that the code is ready, you can run it. You should see output similar to the following screenshot:

You can now track your object.

Objective complete – mini debriefing

Now that you have the code, you can modify the color or add more colors. You also have the location of your object, so later you can attempt to follow the object or manipulate it in some way.

Classified intel

OpenCV is an amazing powerful library of functions. You can do all sorts of incredible things with just a few lines of code. Another common feature you may want to add to your projects is motion detection. If you'd like to try, there are several good tutorials, try looking at:

- http://derek.simkowiak.net/motion-tracking-with-python/
- http://stackoverflow.com/questions/3374828/how-do-i-track-motion-using-opencv-in-python
- https://www.youtube.com/watch?v=8QouvYMfmQo
- https://github.com/RobinDavid/Motion-detection-OpenCV

Mission accomplished

Your projects can now speak and see! You can issue commands, and your projects can respond to changes in the physical environment sensed by the webcam. Next, you will add mobility using motors, servos, and in other ways.

Challenges

Having a webcam connected to your system provides all kinds of additional capabilities. One of the absolute neatest devices out there is Kinect for the Xbox. This device provides not only video, but depth using an infrared device. There are individuals working to make Kinect work with the BeagleBone Black. Several good libraries enable Kinect on Ubuntu. If you'd like to try, buy a used Kinect and then go to `http://speculatrix.tumblr.com/post/23043561344/kinect-on-the-beagleboard-and-ubuntu` or `http://kinepeutics.blogspot.com/2012/04/ethernet-working-installing-kinect.html` and give it a try. Just a word of warning, this task is not for beginners. Later we will talk about the Robot Operating System, which may make it easier.

Also, you can get 3D vision with OpenCV using two cameras. There are several good places for example code, for example in the `samples/cpp` directory that came with OpenCV there is an example `stereo_match.cpp`. Also, for more code examples, you can visit `http://code.google.com/p/opencvstereovision/source/checkout`.

5

Making the Unit Mobile – Controlling Wheeled Movement

You can now talk to the board, and it can talk back. It can even see. Now, you will add the capability to move the entire project using wheels.

Mission briefing

Perhaps the easiest way to make your projects mobile is to add a wheeled platform. In this project, you will be introduced to some of the basics of controlling DC motors and using the BeagleBone Black to control the speed and direction of your wheeled platform.

Why is it awesome?

Even though you can talk to your board, and it can talk back and see, you need to make it mobile to really call it a robot. In this project, you'll learn how to attach your board, both mechanically and electrically, to a wheeled platform. Mobility—what could be more amazing than that?

Downloading the example code and colored images

You can download the example code files and colored images for this Packt book you have purchased from your account at `http://www.packtpub.com`. If you purchased this book elsewhere, you can visit `http://www.packtpub.com/support` and register to have the files e-mailed directly to you.

Your objectives

In this project you will:

- Use a motor controller to control the speed of your platform
- Control your mobile platform programmatically using the BeagleBone Black
- Make your platform truly mobile by issuing voice commands

Mission checklist

You'll need to add some HW, specifically a wheeled or tracked platform, to make your project mobile. There are a lot of choices. Some are completely assembled, others have some assembly required, or you can buy the components and construct your own custom mobile platform. Throughout this book, I'm going to assume that you don't want to do any soldering or mechanical machining yourself, so let's look at a couple of the more popular variants that are available completely assembled or can be assembled with simple tools (screwdriver and/ or pliers). The following is a numbered list of the specific items:

1. The easiest mobile platform is one that has two DC motors that each control a single wheel with a small ball in the front or back. The following is an image of one, sold by SparkFun, called the **Magician Chassis**:

   I did have to assemble this one, but it was fairly straightforward. For more choices in two-wheeled platforms, go to http://www.robotshop.com/2-wheeled-development-platforms-1.html. You could also choose a tracked platform instead of a wheeled platform. A tracked platform has more traction, but is not as nimble in that it takes a longer distance to turn. Again, manufacturers make preassembled units. The following is an image of a preassembled tracked platform, made by Dagu. It's called the **Dagu Rover 5 Tracked Chassis**:

2. Since you have a mobile platform, you'll need a mobile power supply for the BeagleBone Black. I personally like the external 5V rechargeable cell phone batteries that are available almost anywhere that sells mobile phones. These batteries can be charged using a USB cable, either through a DC power supply or directly for a computer USB port. Choose one that comes with two USB output connectors. You'll need them both at some point: one for the BeagleBone Black and one for your powered USB hub. See one in the following image:

You'll also need a USB cable to connect your battery to the BeagleBone Black, but you can just use the cable supplied with the BeagleBone Black.

3. Use your powered USB hub; you'll need to power it not from a wall socket but from the same battery.

4. To power your USB hub, purchase a USB power cable that can connect to your powered hub. My USB hub required a USB to 5.5 mm / 2.1 mm 5V DC Barrel Jack Power cable, which I purchased quite inexpensively from www.amazon.com. The CAT5 cable is the LAN connection. The following image shows how the configuration looks like so that it can be mobile from a power perspective:

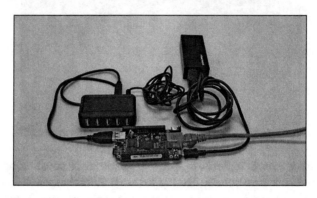

5. Now that you have the mobile platform, you'll need one more bit of HW to connect your BeagleBone Black to it. You'll need some HW that will take the control signals from your BeagleBone Black and turn them into a voltage to control the speed of the motors. Unfortunately, the board cannot source enough current to power the motors directly, so you'll need a circuit that can do this. I strongly suggest you purchase a motor controller instead of designing and building your own. There are numerous possibilities; however, I'm going to suggest one that requires no internal programming and allows you to talk over USB to control the motors. It costs a bit more because you'll need two of them, but it will require much less time and effort to get things rolling. It is a simple motor controller from Pololu: the **Pololu #1372 Simple Motor Controller 18V7 (Fully Assembled)** from www.pololu.com. See its following image:

When you purchase the unit, make sure that you purchase the fully assembled unit; it does come unassembled as well.

This piece of HW will turn USB commands into voltage that controls your motors. You'll need two of these since you are going to control two motors. Also, you are going to connect the controllers to the BeagleBone Black via USB, so you'll need the USB hub and two USB A to mini-B cables.

6. The final piece you will need is two short pieces of wire, less than 2 inches, stripped at the end. You can buy these wires, called **male-male jumper wires**, online, for example from `www.pololu.com` or `www.amazon.com`. They also come in male-female and female-male versions. The following is an image of a set I purchased recently:

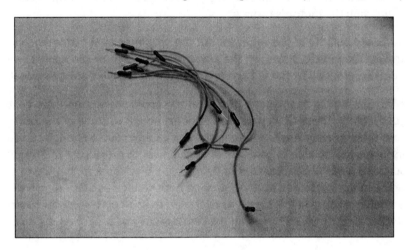

Again, I've selected components to avoid required soldering.

Now that you have all the HW, let's walk through a quick tutorial about how the system works and then some step-by-step instructions to make your project mobile.

Using a motor controller to control the speed of your platform

The first step to make the platform mobile is adding a motor controller. This allows us to control the speed of each wheel (or track) independently.

Prepare for lift off

Before you get started, let's spend some time understanding the basics of the motor control. Whether you chose the two-wheeled mobile platform or the tracked platform, the basic movement control is the same. The unit moves by engaging the motors. If the desired direction is straight, the motors run at the same speed. If you want to turn the unit, the motors run at different speeds. The unit can turn in a circle if you run one motor forward and one backwards.

DC motors are fairly straightforward devices. The speed and direction of the motor is controlled by the magnitude and polarity of the voltage applied to its terminals. The higher the voltage, the faster the motor will turn. If you reverse the polarity of the voltage, you can reverse the direction the motor is turning.

The magnitude and polarity of the voltage is not the only important factor when you think about controlling the motors, however; the power that your motor can apply to moving your platform is also determined by the voltage and the current supplied at its terminals.

There are GPIO pins on the BeagleBone Black that you could use to create the control voltage and drive your motors. These GPIO pins provide direct access to some of the control lines available from the processor itself. However, the unit cannot source enough current, and your motors would not be able to generate enough power to move your mobile platform. You can also cause physical damage to your BeagleBone Black board. That is why you will need to use the motor controller, as it will provide both voltage and current, so that your platform can move reliably. In this case, I have chosen to hook up two motor controllers that can be controlled directly via USB, making the connections and programming much simpler.

Engage thrusters

The first step in making your project mobile is connecting the motor controller to the platform. There are two connections you need to make. First, you'll need to connect a battery to the controller. Second, you'll need to connect the motors themselves.

To connect the battery, find the output connectors on the battery holder. On the wheeled platform, you'll need to do a little work, as the connector on the battery has a plug. Remove the plug and strip back the wire about a half an inch. It should look like the following image:

On the back of the motor controller, notice the labels **VIN**, **OUTB**, **OUTA**, and **GND**, which are shown in the following image:

Once you have the battery pack ready, insert the wires into the motor controller in the blue connectors marked **VIN** and **GND**. **VIN** is the constant DC voltage in from your batteries, and **GND** is the ground connection from your batteries. **OUTA** and **OUTB** are the control signals to your DC motors. You'll notice that close to the battery connector, one of the wires is red. Follow that wire to the end and insert it into the **VIN** connector, and then tighten the screw connector. Close to the battery connector, you'll notice one of the wires is black: insert that wire into the connector marked **GND** and tighten the screw connector.

The package will look like the following image:

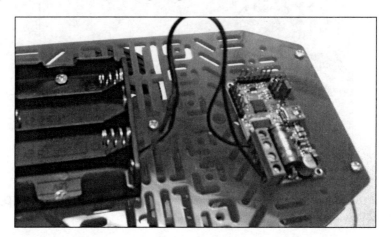

Now, connect one of the motors to the motor controller by connecting the red and black wires with male connectors to the two inner blue screw connectors, the red one to **OUTA** and the black one to **OUTB**. In order to test the platform, turn it over. The entire system should now look like the following image:

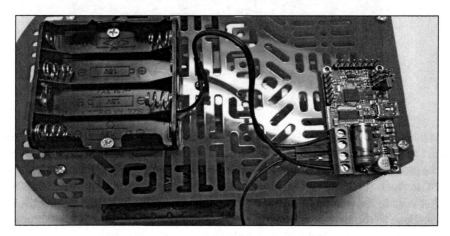

You can now test the system by using SW provided by Pololu. If you want to first try it on a Windows PC, install the Windows driver SW from `http://www.pololu.com/docs/0J44/3.1`. Unzip and install the SW, then connect your motor controller to the PC using a USB cable and then start the SW. You should see a screen as shown in the following screenshot:

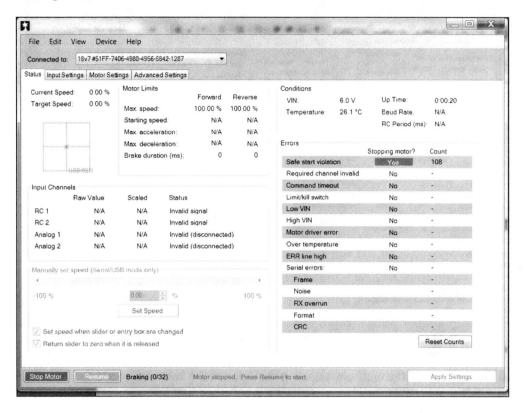

The **Safe start violation** tab is set when you first enter the program; you need to clear this by clicking on the **Resume** button at the bottom-left corner of the screen. The screen should now look like the following screenshot:

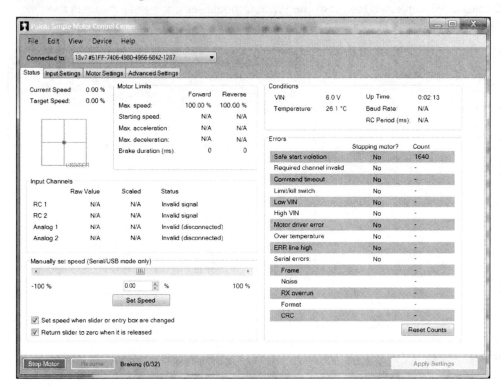

You can now control the motor by selecting the slider on the lower-left section of the screen. The motor should now turn when you select a value other than zero percent, as shown in the following screenshot:

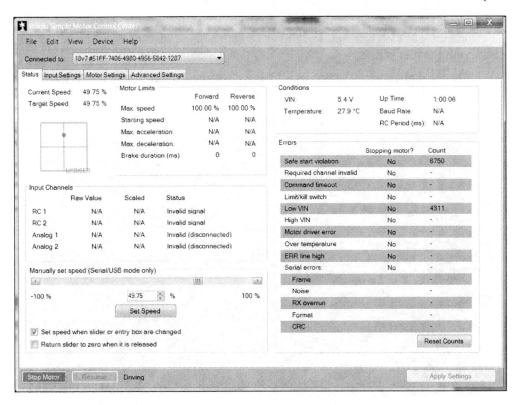

The motor should now be turning! You have control of the DC motor. Repeat the previous steps with the second controller.

You can also do some limited direct control of your motor controller through your BeagleBone Black. If you are going to do this remotely, log on through PuTTY. You'll also want to open a vncview session to have access to a web browser. If you are doing this directly on a monitor, simply log in. Then, perform the following steps:

1. Open a web browser window and type the URL of the Linux version of your motor controller from `http://www.pololu.com/docs/0J44/3.2`. This is the Linux version of your test SW. Download this code by selecting the link on the web page and selecting the **Save File** button.

2. Move the code to the home directory by going to the `Download` directory and using the `mv` command by typing: `mv smc-linux-101119.tar.gz`. (The specific number may be slightly different, as revisions change. Only one version will be available on the website; use that version.)

3. Unzip the file you just moved by using the command: `tar -xzvf smc-linux-101119.tar.gz`, and this will create a new directory `smc-linux` with the files in it.

4. Change directory (`cd`) to the `smc-linux` directory and list the files (`11`). You should see something like the following screenshot:

5. The following are two steps defined in the `README.txt` file that you need to perform:

   ❑ First, you need to download some additional SW. To do this, type: `sudo apt-get install libusb-1.0-0-dev mono-runtime libmono-winforms2.0-cil`.

   ❑ Second, you need to copy a file in order to give yourself privileges to access the HW. To do this, type `sudo cp 99-pololu.rules /etc/udev/rules.d/` and all users will be able to access the HW.

6. Now, you should plug your motor controller into the powered USB hub; then plug the powered USB into the to the BeagleBone Black. The configuration will look like the following image:

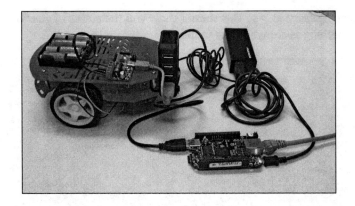

7. You'll need to do a `sudo reboot` on the BeagleBone Black to get it to recognize the motor controller. Once you have done this, either log in directly or open a PuTTY window. Unfortunately, you can't run the graphical user interface program on your BeagleBone Black, so you won't need a vncview session, but you can run the `SmcCmd` program, which will allow you to configure and control the motor controller. Type `./SmcCmd`, and you will get the list of possible interactions with the motor controller. They will look as shown in the following screenshot:

```
ubuntu@ubuntu-armhf:~/smc_linux$ ./SmcCmd
SmcCmd: Configuration and control utility for the Simple Motor Controller.
Version: 1.1.0.0
Options:
 -l, --list                 list available devices
 -d, --device SERIALNUM     (optional) select device with given serial number
 -s, --status               display complete device status
     --stop                 stop the motor
     --resume               allow motor to start
     --speed NUM            set motor speed (-3200 to 3200)
     --brake NUM            stop motor with variable braking.  32=full brake
     --restoredefaults      restore factory settings
     --configure FILE       load settings file into device
     --getconf FILE         read device settings and write to file
     --bootloader           put device in bootloader (firmware upgrade) mode
Options for changing motor limits until next reset:
     --max-speed NUM         (3200 means no limit)
     --max-speed-forward NUM (3200 means no limit)
     --max-speed-reverse NUM (3200 means no limit)
     --max-accel NUM
     --max-accel-forward NUM
     --max-accel-reverse NUM
     --max-decel NUM
     --max-decel-forward NUM
     --max-decel-reverse NUM
     --brake-dur NUM         units are ms.  rounds up to nearest 4 ms
     --brake-dur-forward NUM units are ms.  rounds up to nearest 4 ms
     --brake-dur-reverse NUM units are ms.  rounds up to nearest 4 ms
ubuntu@ubuntu-armhf:~/smc_linux$
```

8. First, type `./SmcCmd -s`. This should show you the status of your device. It should look something as shown in the following screenshot:

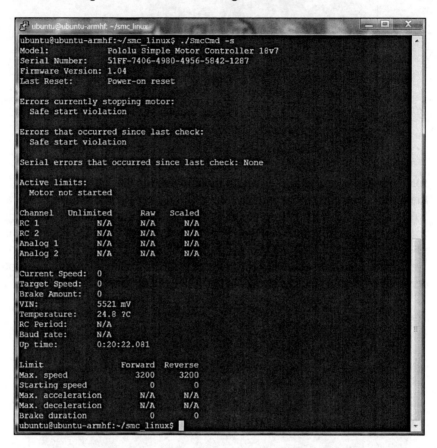

9. Now, you can issue commands to your motor controller. First, you need to clear the errors currently stopping the motor, in this case the default **Safe start violation**. Type the command: `./SmcCmd --resume`. Then issue the command: `./SmcCmd --speed 1000`, and the motor should be turning! You can turn the motor off by issuing the command: `./SmcCmd --stop`.

Objective complete – mini debriefing

Now that you have your motor running, your next step is to plug both motor controllers into the USB hub and control both motors programmatically using the BeagleBone Black.

Controlling your mobile platform programmatically using the BeagleBone Black

Now that you have your basic motor controller functionality up and running, you need to connect both motor controllers to the BeagleBone Black. This task will cover this, and then show you how to control your entire platform programmatically.

Prepare for lift off

Now you'll hook up both motor controllers to the battery power supply and motors. Let's start with the motor controllers. The screw-type connectors on your motor controllers make this easy and help you avoid any soldering.

First, run the battery connectors to the **VIN** and **GND** connections on one of the motor controllers. Then, take one of the two small lengths of the jumper wire and place that in the **VIN** connector, and then do the same with the second jumper wire and the **GND** connector, as shown in the following image:

Now screw the connection tight. Do this on both the **VIN** and **GND** connectors. Now, take the other end of these two jumper wires and place them in the **VIN** and **GND** connections on the other motor controller and screw them tight. Now, you have power connections to both of your motor controllers.

The next step is to connect each of the motors to one of the motor controllers, using the same technique that you used in the last task. Take the red and black wire connections from the motor and place them in the **OUTA** and **OUTB** connections, the red in the **OUTA** connection and the black in the **OUTB** connection, as shown in the following image:

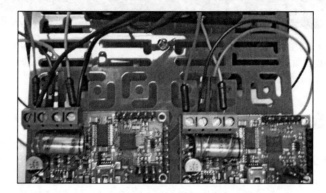

Now, you need to connect the BeagleBone Black and its USB hub to the motor controllers. First, connect the USB to the motor controllers using the USB cables. Then, connect the USB hub to the BeagleBone Black using its USB cable. These connections are shown in the following image:

Once you've made these connections, you can configure all of the HW on top of the mobile platform, perhaps as shown in the following image:

I tend to use lots of cable ties, but if you'd like it to look even more polished, feel free to use more metal nuts and bolts.

Engage thrusters

I suggest you use Python in your initial attempts to control the motor. It is very straightforward to code, run, and debug your code in Python. I am going to include the directions in this task for Python; you can also go to the Pololu's website at `www.pololu.com/` and find instructions for how to access the capabilities using C.

The first Python program you are going to create is shown in the following screenshot:

```
#!/usr/bin/python

import serial
import time
class MotorControllerOne(object):
    def __init__(self, port= "/dev/ttyACM0"):
        self.ser = serial.Serial(port = port)
    def exitSafeStart(self):
        command = chr(0x83)
        self.ser.write(command)
        self.ser.flush()
    def setSpeed(self, speed):
        if speed > 0:
            channelByte = chr(0x85)
        else:
            channelByte = chr(0x86)
        lowTargetByte = chr(speed & 0x1F)
        highTargetByte = chr((speed >> 5) & 0x7F)
        command = channelByte + lowTargetByte + highTargetByte

        self.ser.write(command)
        self.ser.flush()
    def close(self):
        self.ser.close()

if __name__ =="__main__":

    motor1 = MotorControllerOne()
    motor1.exitSafeStart()
    time.sleep(.2)
    motor1.setSpeed(int(2000))
    time.sleep(1)
    motor1.setSpeed(int(0))
    time.sleep(1)
```

To create this program, go to your `smc_linux` directory and type `emacs dcmotor.py` (if you are using a different editor, open a new file with the name `dcmotor.py`). Now enter the program. Let's go through the program step by step:

1. The first line allows your program to run outside of the Python environment. You'll use it later when you want to execute your code using voice commands.

2. The next line imports the serial library. You need this to talk to our motor controllers.

3. The `MotorControllerOne` class holds four functions. The `__init__` function associates your motor controller with the specific serial port, in this case `ttyACM0`. The `exitSafeStart` function tells your motor that you want to actually run it now and removes the safe start setting that comes as default. The `setSpeed` function takes your `speed` setting and turns it into a serial command that the unit can understand, and then sends the command. The `close` function closes the port when you leave your program.

4. The `if __name__=="__main__"` section is the main part of your program. The first line here initializes your motor controller; the second line tells your motor controller to exit the safe start default. The third line is 200 milliseconds wait, then the fourth line tells your motor to turn at a speed of `2000`. The fourth line in this section waits one second then the fifth line tells your motor to go back to the `0` speed. The final line is a one second wait.

5. In order to run this program, you'll need the serial library. Install it by typing `sudo apt-get install Python-serial` at the prompt. You'll then need to add yourself to the dialout group; do this by typing `sudo adduser ubuntu dialout`. Then, do a `sudo reboot` to enable all these changes.

6. With this installed, you can run your program. To do this, type `Python dcmotor.py`. Your motor should run for one second then stop. You can now control the motor through Python! Additionally, you'll want to make this program available to run from the command line. Type: `chmod +x dcmotor.py`. If you now do an `ll` (list programs), you'll see that your program is now green, which means you can execute it directly. Now, you can type `./dcmotor.py`.

7. The final step is to create a second controller for the second motor. Do this, by adding a `MotorControllerTwo` class that is a copy of the `MotorControllerOne` class except that the port it points to is `ttyACM1`. This code will look as shown in the following screenshot:

```
                self.ser.close()

class MotorControllerTwo(object):
    def __init__(self, port= "/dev/ttyACM1"):
        self.ser = serial.Serial(port = port)
    def exitSafeStart(self):
        command = chr(0x83)
        self.ser.write(command)
        self.ser.flush()
    def setSpeed(self, speed):
        if speed > 0:
            channelByte = chr(0x85)
        else:
            channelByte = chr(0x86)
        lowTargetByte = chr(speed & 0x1F)
        highTargetByte = chr((speed >> 5) & 0x7F)
        command = channelByte + lowTargetByte + highTargetByte

        self.ser.write(command)
        self.ser.flush()
    def close(self):
        self.ser.close()

if __name__=="__main__":

    motor1 = MotorControllerOne()
    motor2 = MotorControllerTwo()
    motor1.exitSafeStart()
    motor2.exitSafeStart()
    time.sleep(.2)
    motor1.setSpeed(int(2000))
    motor2.setSpeed(int(2000))
    time.sleep(1)
    motor1.setSpeed(int(0))
    motor2.setSpeed(int(0))
    time.sleep(1)
-=--:**--F1  dcmotor.py      Bot L42    (Python)-------------------------------
```

You will also copy the statements in the main section, so that both `motor1` and `motor2` do the same things. You don't need to copy the `time.sleep` statements. They are just fixed delays. Now when you run your program, both motors should turn. One important note; Linux is not a real-time platform, so your motors cannot be guaranteed to turn at exactly the same time. However, they are normally going to move within a few milliseconds of each other, which is good enough in this application. My platform turned out to be a bit finicky, I found I had to do a `sudo reboot` and re-log in using Putty to reset the USB, so that I could run the program more than once. Also, you may need to issue the `./SmcCmd --resume` command to reset the motor controllers. A bit painful, but it works.

Objective complete – mini debriefing

Now that you know the basics of commanding your mobile platform, feel free to add even more `setSpeed` commands to make your mobile platform move. Setting both the motors to a positive speed will move the mobile platform forward. Setting them to a negative value will make the platform go in reverse. Running just one motor will make the platform turn, as will running both the motors in opposite directions.

Classified intel

The platforms you've looked at up until now had two DC motors. It would be easy to add even more motors. There are several platforms that provide DC motors for all four wheels. In this case, you'd just add two more motor controllers, and then update the code for four classes of `MotorContoller`.

Making your mobile platform truly mobile by issuing voice commands

Prepare for lift off

You should now have a mobile platform that you can program to move in any number of ways. Unfortunately, you still have your LAN cable connected, so the platform isn't completely mobile. And once you have begun the program, you can't alter the behavior of your program. In this task, you will use the principles from *Chapter 2*, *Programming the BeagleBone Black*, to issue voice commands to initiate movement.

Engage thrusters

You'll need to modify your voice recognition program, so it will run your Python program when it gets a voice command. If you feel rusty on how this works, review *Chapter 2*, *Programming the BeagleBone Black*. You are going to make a simple modification to the `continuous.c` program in `/home/ubuntu/pocketsphinx-0.8/src/programs`. To do this, type: `cd /home/ubuntu/pocketsphinx-0.8/src/programs` and then type `emacs continuous.c`. The changes will come in the same section as your other voice commands, and will look as shown in the following screenshot:

```
ubuntu@ubuntu-armhf: ~/pocketsphinx-0.8/src/programs
File Edit Options Buffers Tools C Help
        printf("Stopped listening, please wait...\n");
        fflush(stdout);
        /* Finish decoding, obtain and print result */
        ps_end_utt(ps);
        hyp = ps_get_hyp(ps, NULL, &uttid);
        printf("%s: %s\n", uttid, hyp);
        fflush(stdout);

        /* Exit if the first word spoken was GOODBYE */
        if (hyp) {
            sscanf(hyp, "%s", word);
            if (strcmp(word, "GOOD BYE") == 0)
                {
                    system("espeak \"good bye\"");
                    break;
                }
            else if (strcmp(hyp, "HELLO") == 0)
                system("espeak \"hello\"");
            else if (strcmp(hyp, "FORWARD"))
                {
                    system("espeak \"moving robot\"");
                    system("/home/ubuntu/smc_linux/dcmotor.py");
                }
        }
-==-:----F1  continuous.c   78% L328   (C/l Abbrev)-----------------
Wrote /home/ubuntu/pocketsphinx-0.8/src/programs/continuous.c
```

The additions are pretty straightforward. Let's walk through them:

1. `else if (strcmp(hyp, "FORWARD") == 0)`: This checks the word as recognized by your voice command program. If it corresponds with the word `FORWARD`, you will execute everything inside the `if` statement. You use `{ }` to group and tell the system which commands go with this `else if` clause.

2. `system("espeak \"moving robot\"")`: This executes `espeak`, which should tell us that you are about to run your robot program. By the way, you need to type `\"` because the `"` character is a special character in Linux, and if you want the actual `"` character, you need to proceed it with the `\` character.

3. `system("/home/ubuntu/smc_linux/dcmotor.py")`: This is the program you will execute. In this case, your mobile platform will do whatever the `dcmotor.py` program tells it to do.

After doing this, you will need to recompile the program, so type `make` and the executable `pocketsphinx_continuous` will be created. Run the program by typing `./pocketsphinx_continuous`. Disconnect the LAN cable, and the mobile platform will now take the "forward" voice command and execute your program.

Objective complete – mini debriefing

You should now have a complete mobile platform! When you execute your program, the mobile platform can now move around based on what you have programmed it to do.

Classified intel

You don't have to put all of your capabilities in one program. You can create several programs, each with a different function, and then connect each of the programs to their appropriate voice commands. Perhaps one program moves your robot forward, a different backwards, another to turn right or left.

Mission accomplished

Now you have your mobile platform up and ready to move around. You can command it using your voice. In the next project, you'll introduce a different kind of mobile platform: one with legs.

A challenge

You have already covered how to add vision to your BeagleBone Black project. A great addition to your mobile robot is the ability to follow a colored object attached to a target.

Remember how you used OpenCV to find a colored object, and could then find out where in your field of view (left or right, or up or down) it existed? You can use this to decide whether to move your mobile platform right or left, or forward or backward. Try this, and then move the target to see if your mobile robot can follow it.

6

Making the Unit Very Mobile – Controlling Legged Movement

In the previous chapter, we covered wheeled and tracked movement. Cool enough, but what if you want your robot to navigate uneven ground? Now you will add the capability to move the entire project using legs.

Mission briefing

We've covered creating robots using a wheeled/track base. In this chapter, you will be introduced to some of the basics of servo motors and using the BeagleBone Black to control the speed and direction of your legged platform. Here is an image of a finished project:

Why is it awesome?

Even though you've learned to make your robot mobile by adding wheels or tracks, this mobile platform will only work well on smooth, flat surfaces. Often, you'll want your robot to work in environments where it is not smooth or flat; perhaps, you'll even want your robot to go upstairs or over curbs. In this chapter, you'll learn how to attach your board, both mechanically and electrically, to a platform with legs, so your projects can be mobile in many more environments. Robots that can walk: what could be more amazing than that?

Your objectives

In this chapter, you will learn:

▶ Connecting the BeagleBone Black to a mobile platform using a servo controller

▶ Creating a program in Linux to control the movement of the mobile platform

▶ Making your mobile platform truly mobile by issuing voice commands

Downloading the example code and colored images

You can download the example code files and colored images for this Packt book you have purchased from your account at `http://www.packtpub.com`. If you purchased this book elsewhere, you can visit `http://www.packtpub.com/support` and register to have the files e-mailed directly to you.

Mission checklist

In this chapter, you'll need to add a legged platform to make your project mobile. So, here is your parts' list:

▶ **A legged robot**: There are a lot of choices. As before, some are completely assembled, others have some assembly required, and you may even choose to buy the components and construct your own custom mobile platform. Also, as before, I'm going to assume that you don't want to do any soldering or mechanical machining yourself, so let's look at a several choices that are available completely assembled or can be assembled by simple tools (screwdriver and/or pliers).

One of the easiest legged mobile platforms is one that has two legs and four servo motors. Here is an image of this type of platform:

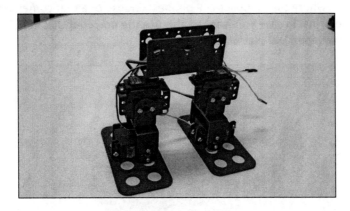

You'll use this platform in this chapter because it is the simplest to program and because it is the least expensive, requiring only four servos. To construct this platform, you must purchase the parts and then assemble it yourself. Find the instructions and parts list at `http://www.lynxmotion.com/images/html/build112.htm`. Another easy way to get all the mechanical parts (except servos) is to purchase a biped robot kit with six **degrees of freedom** (**DOF**). This will contain the parts needed to construct your four-servo biped. These six DOF bipeds can be purchased by searching eBay or by going to `http://www.robotshop.com/2-wheeled-development-platforms-1.html`.

▶ You'll also need to purchase the **servo motors**. For this type of robot, you can use standard size servos. I like the Hitec HS-311 or HS-322 for this robot. They are inexpensive but powerful enough. You can get those on Amazon or eBay. Here is an image of an HS-311:

▸ As in the last chapter, you'll need a **mobile power supply** for the BeagleBone Black. Again, I personally like the 5V cell phone rechargeable batteries that are available almost anywhere that supplies cell phones. Choose one that comes with two USB connectors, just in case you want to also use the powered USB hub. This one mounts well on the biped HW platform:

▸ You'll also need a **USB cable** to connect your battery to the BeagleBone Black, but you can just use the cable supplied with the BeagleBone Black. If you want to connect your powered USB hub, you'll need a USB to DC jack adapter for that as well.

▸ You'll also need a way to connect your batteries to the servo motor controller. Here is an image of a **four AA battery holder**, available at most electronics parts stores or from Amazon:

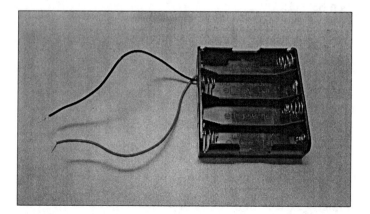

- Now that you have the mechanical parts for your legged mobile platform, you'll need some HW that will take the control signals from your BeagleBone Black and turn them into a voltage that can control the servo motors. Servo motors are controlled using a control signal called **PWM**. For a good overview of this type of control, see `http://pcbheaven.com/wikipages/How_RC_Servos_Works/` or `https://www.ghielectronics.com/docs/18/pwm`. You can find tutorials that show you how to control servos directly using the BeagleBone Black's GPIO pins, for example, here at `http://learn.adafruit.com/controlling-a-servo-with-a-beaglebone-black/overview` and `http://www.youtube.com/watch?v=6gv3gWtoBWQ`. For ease of use I chose to purchase a **motor controller** that can talk over USB and control the servo motor. These protect my board and make controlling many servos easy. My personal favorite for this application is a simple servo motor controller utilizing USB from Pololu that can control 18 servo motors. Here is an image of the unit:

Again, make sure you order the assembled version. This piece of HW will turn USB commands into voltage that control your servo motors. Pololu makes a number of different versions of this controller, each able to control a certain number of servos. Once you've chosen your legged platform, simply count the number of servos you need to control, and chose the controller that can control that number of servos. One advantage of the 18 servo controller is the ease of connecting power to the unit via screw type connectors.

- Since you are going to connect this controller to your BeagleBone Black via USB, you'll also need a **USB A to mini-B cable**.

Now that you have all the HW, let's walk through a quick tutorial on how a two-legged system with servos works and then some step-by-step instructions to make your project walk.

Connecting the BeagleBone Black to the mobile platform using a servo controller

Now that you have a legged platform and a servo motor controller, you are ready to make your project walk!

Prepare for lift off

Before you begin, you'll need some background on servo motors. Servo motors are somewhat similar to DC motors; however, there is an important difference. While DC motors are generally designed to move in a continuous way—rotating 360 degrees at a given speed—servos are generally designed to move within a limited set of angles. In other words, in the DC motor world, you generally want your motors to spin with continuous rotation speed that you control. In the servo world, you want your motor to move to a specific position that you control.

Engage thrusters

To make your project walk, you first need to connect the servo motor controller to the servos. There are two connections you need to make: the first to the servo motors, the second to the battery holder. In this section, you'll connect your servo controller to your PC to check to see if everything is working.

1. First, connect the servos to the controller. Here is an image of your two-legged robot, and the four different servo connections:

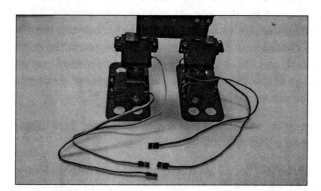

2. In order to be consistent, let's connect your four servos to the connections marked 0 through 3 on the controller using this configuration: 0 – left foot, 1 – left hip, 2 – right foot, and 3 – right hip. Here is an image of the back of the controller; it will tell you where to connect your servos:

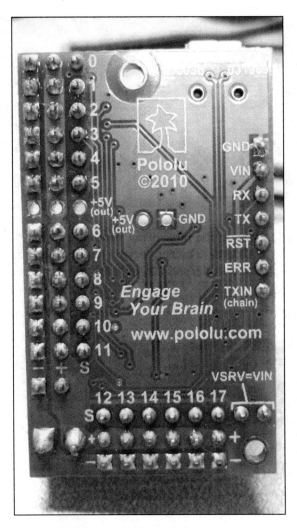

3. Connect these to the servo motor controller like this: the left foot to the top O connector, black cable to the outside (–), the left hip to the 1 connector, black cable out, right foot to the 2 connector, black cable out, and right hip to the 3 connector, black cable out. See the following image for a clearer description:

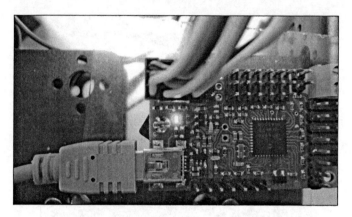

4. Now you need to connect the servo motor controller to your battery. If you are using a standard 4 AA battery holder, connect it to the two green screw connectors, the black cable to the outside, and the red cable to the inside, as shown in the following image:

5. Now you can connect the motor controller to your PC to see if you can talk with it.

Objective complete – mini debriefing

Now that the HW is connected, you can use some SW provided by Polulu to control the servos. It is easiest to do this using your personal computer. First, download the Polulu SW from `http://www.pololu.com/docs/0J40/3.a` and install it based on the instructions on the website. Once it is installed, run the SW, and you should see the following screen:

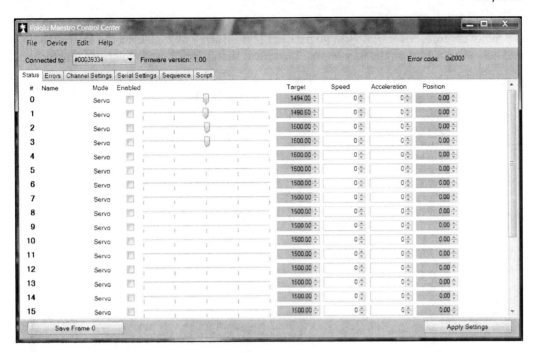

You first will need to change the configuration on **Serial Settings**, so select the **Serial Settings** tab, and you should see a screen as shown in the following screenshot:

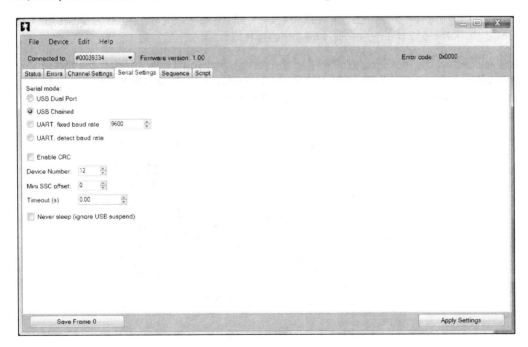

Make sure that the **USB Chained** option is selected; this will allow you to connect and control the motor controller over USB. Now go back to the main screen by selecting the **Status** tab, and now you can turn on the four servos. The screen should look like the following screenshot:

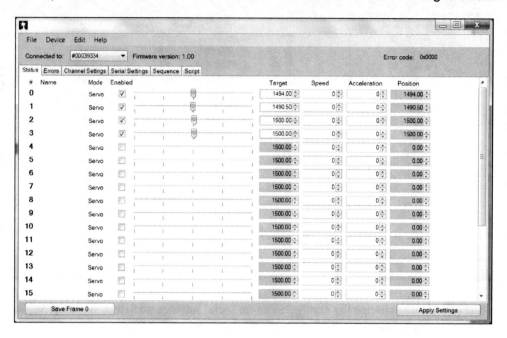

Now you can use the sliders to control the servos. Make sure that the servo 0 moves the left foot, 1 the left hip, 2 the right foot, and 3 the right hip.

You've checked the motor controllers and the servos, and you'll now connect the motor controller up to the BeagleBone Black control the servos from it. Remove the USB cable from the PC and connect it into the powered USB hub. The entire system will look like the following image:

Let's now talk to the motor controller by downloading the Linux code from Pololu at `http://www.pololu.com/docs/0J40/3.b`. Perhaps, the best way is to log in to your Beagle Bone Black by using **vncserver** and a **vncviewer** window on your PC. To do this, log in to your BeagleBone Black using PuTTY, then type `vncserver` at the prompt to make sure vncserver is running.

1. On your PC open the VNC Viewer application, enter your IP address, then press **connect**. Then enter your password that you created for the vncserver, and you should see the BeagleBone Black Viewer screen, which should look like this:

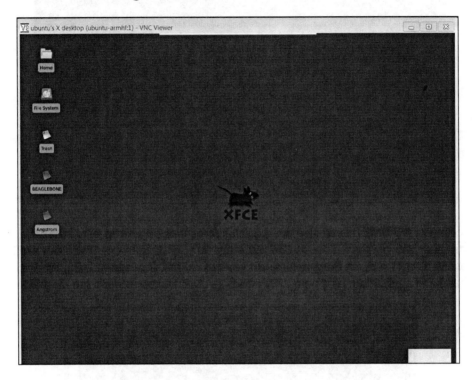

2. Open a Firefox browser window and go to `http://www.pololu.com/docs/0J40/3.b`. *Click* on the Maestro Servo Controller Linux Software link. You will download the file **maestro_linux_100507.tar.gz** to the **Download** directory.

3. Go to your download directory, move this file to your home directory by typing `mv maestro_linux_100507.tar.gz ..` and then you can go back to your home directory.

4. Unpack the file by typing `tar -xzfv maestro_linux_011507.tar.gz`. This will create a directory called **maestro_linux**. Go to that directory by typing `cd maestro_linux` and then type `ls`. You should see something like this:

```
ubuntu@ubuntu-armhf:~$ ls
Desktop     Music      Templates     maestro_linux_100507.tar.gz
Documents   Pictures   Videos        smc-linux-101119.tar.gz
Downloads   Public     maestro_linux smc_linux
ubuntu@ubuntu-armhf:~$ cd maestro_linux
ubuntu@ubuntu-armhf:~/maestro_linux$ ls
99-pololu.rules  FirmwareUpgrade.dll   README.txt        UsbWrapper.dll  UscCmd
Bytecode.dll     MaestroControlCenter  Sequencer.dll     Usc.dll
ubuntu@ubuntu-armhf:~/maestro_linux$
```

The document **README.txt** will give you explicit instructions on how to install the SW. Unfortunately you can't run **MaestroControlCenter** on your BeagleBone Black. Our version of windowing doesn't support the graphics, but you can control your servos using the **UscCmd** command-line application. First type `./UscCmd --list` and you should see the following:

```
ubuntu@ubuntu-armhf:~$ ls
Desktop     Music      Templates     maestro_linux_100507.tar.gz
Documents   Pictures   Videos        smc-linux-101119.tar.gz
Downloads   Public     maestro_linux smc_linux
ubuntu@ubuntu-armhf:~$ cd maestro_linux
ubuntu@ubuntu-armhf:~/maestro_linux$ ls
99-pololu.rules  FirmwareUpgrade.dll   README.txt        UsbWrapper.dll  UscCmd
Bytecode.dll     MaestroControlCenter  Sequencer.dll     Usc.dll
ubuntu@ubuntu-armhf:~/maestro_linux$ ./UscCmd --list
1 Maestro USB servo controller device found:
#00039334
ubuntu@ubuntu-armhf:~/maestro_linux$
```

The unit sees your servo controller. If you just type ./UscCmd you can see all the commands you could send to your controller:

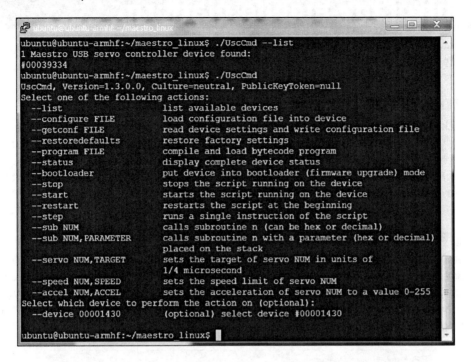

Notice you can send a servo a specific target angle, although the target is not in angle values, so it makes it a bit difficult to know where you are sending your servo. Try typing ./UscCmd --servo 0, 10. The servo will most likely move to its full angle position. Type ./UscCmd --servo 0, 0 and it will stop the servo from trying to move. In the next section, you'll write some SW that will translate your angles to the commands that the servo controller will want to see.

If you didn't run the Windows version of Maestro Controller and set the serial settings to **USB Chained**, your motor controller may not respond.

Creating a program in Linux to control the mobile platform

Now that you can control your servos using basic commands, let's control them using a program.

Prepare for lift off

So, you know that you can talk to your servo motor controller and set your servos. In this section, you'll create a Python SW program that will let you talk to your servos a bit more intuitively. You'll issue commands that tell a servo to go to a specific angle, and it will go to that angle. You can then add a set of such commands to allow your legged mobile robot to lean left, lean right, or even take a step forward.

Engage thrusters

Let's start with a simple program that will make your legged mobile robot's servos go to 90 degrees. This should be somewhere close to the middle of the 180 degrees you can set. However, the center, maximum, and minimum values can vary from servo to servo, so you may need to calibrate these values. To keep things simple, we will not cover that here. The following is the code:

```
ubuntu@ubuntu-armhf: ~/maestro_linux
File Edit Options Buffers Tools Python Help
#!/usr/bin/python
import serial
import time
class PololuMicroMaestro(object):
    def __init__(self, port= "/dev/ttyACM0"):
        self.ser = serial.Serial(port = port)
    def setAngle(self, channel, angle):
        minAngle = 0.0
        maxAngle = 180.0
        minTarget = 256.0
        maxTarget = 13120.0
        scaledValue = int((angle / ((maxAngle - minAngle) / (maxTarget - minTar\
get))) + minTarget)
        commandByte = chr(0x84)
        channelByte = chr(channel)
        lowTargetByte = chr(scaledValue & 0x7F)
        highTargetByte = chr((scaledValue >> 7) & 0x7F)
        command = commandByte + channelByte + lowTargetByte + highTargetByte
        self.ser.write(command)
        self.ser.flush()
    def close(self):
        self.ser.close()
if __name__ =="__main__":
    robot = PololuMicroMaestro()
# Home position
    robot.setAngle(0,85)
    robot.setAngle(1,80)
    robot.setAngle(2,80)
    robot.setAngle(3,75)
-=--:----F1  robot.py      Top L1    (Python)-------------------------------
For information about GNU Emacs and the GNU system, type C-h C-a.
```

The explanation of the code is as follows:

▸ `#!/user/bin/python`: This first line allows you to make this Python file execute from the command line. This will allow you to call this program from your voice command program. We'll talk about that in the next section.

▸ `import serial`, `import time`: These next two lines include the `serial` and `time` libraries. You need the `serial` library to talk to your unit via USB, and the `time` library you will use later to wait between servo commands.

▸ The `PololuMicroMaestro` class holds the methods that will allow you to communicate with your motor controller.

▸ The first method, the `__init__` method, opens the USB port associated with your servo motor controller.

▸ The next method, `setAngle`, converts your desired setting of servo and angle into the serial command that the servo motor controller needs. The values, such as `minTarget`, `maxTarget`, and the structure of the communications, `channelByte`, `commandByte lowTargetByte`, and `highTargetByte`, come from the manufacturer.

▸ The last method, `close`, closes the serial port.

▸ Now that you have the class, the `__main__` part of the program instantiates an instance of your servo motor controller class so you can call it.

▸ Now you can set each servo to the desired position. The default would be to set each servo to 90 degrees. However, the servos were not exactly centered, so I found on my robot that I needed to set each servo to the values shown in this program to be lined up so my robot had both feet on the ground and both hips centered.

Once you have the basic home position set, you can now ask your robot to do some things. Here are some examples in simple Python code:

```
ubuntu@ubuntu-armhf: ~/maestro_linux
File Edit Options Buffers Tools Python Help
# Home position
    robot.setAngle(0,85)
    robot.setAngle(1,80)
    robot.setAngle(2,80)
    robot.setAngle(3,75)
    time.sleep(2)
#Lean Right
    robot.setAngle(2,90)
    robot.setAngle(0,110)
    time.sleep(2)
#Lean Left
    robot.setAngle(0,70)
    robot.setAngle(2,60)
    time.sleep(2)
#Step Forward Left
    robot.setAngle(2,90)
    robot.setAngle(0,110)
    time.sleep(.5)
    robot.setAngle(3,100)
    time.sleep(.2)
    robot.setAngle(1,100)
    time.sleep(2)
#Step Forward Right
    robot.setAngle(0,70)
    robot.setAngle(2,60)
    time.sleep(.5)
    robot.setAngle(1,50)
    time.sleep(.2)
    robot.setAngle(3,50)
    time.sleep(2)
#Back to Home
    robot.setAngle(0,85)
    robot.setAngle(1,80)
    robot.setAngle(2,80)
    robot.setAngle(3,75)
```

In this case, you are using your `setAngle` command to set your servos to manipulate your robot. This set of commands first sets your robot to the home position. Then, you can use the feet to lean to the right, then to the left, then you can use a combination of commands to make your robot step forward with the left foot, and then with the right foot.

Objective complete – mini debriefing

Once you have the program working, you'll want to package all your HW onto the mobile robot. There is no right or wrong way to do this, but I like to use a small piece of transparent plastic because it is easy to cut and drill. Here is what my robot looks like:

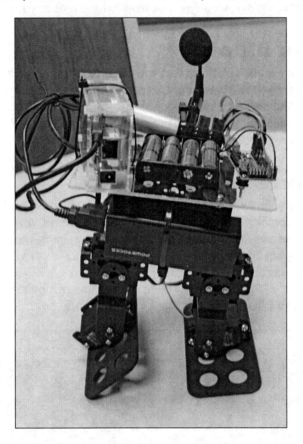

Classified intel

By following these principles, you can make your robot do many amazing things. Walk forward, walk backward, dance, turn around—any number of movements are possible. The best way to learn is to try new and different positions with the servos.

Making your mobile platform truly mobile by issuing voice commands

Now that your robot can move, wouldn't it be neat to have it obey your commands?

Prepare for lift off

You should now have a mobile platform that you can program to move in any number of ways. Unfortunately, you still have your LAN cable connected, so the platform isn't completely mobile. And once you have begun the program, you can't alter the behavior of your program. In this section, you will use the principles from *Chapter 3, Providing Speech Input and Output*, to issue voice commands to initiate movement.

Engage thrusters

You'll need to modify your voice recognition program, so it will run your Python program when it gets a voice command. If you feel rusty on how this works, review *Chapter 3, Providing Speech Input and Output*. You are going to make a simple modification to the `continuous.c` program in `/home/ubuntu/pocketsphinx-0.8/src/programs`. To do this, type `cd /home/ubuntu/ pocketsphinx-0.8/src/programs` and then type `emacs continuous.c`. The changes will come in the same section as your other voice commands, and will look like this:

The additions are pretty straightforward. Let's walk through them:

- ▶ `else if (strcmp(hyp, "FORWARD") == 0)`: This checks the word as recognized by your voice command program. If it corresponds with the word `FORWARD`, you will execute everything inside the `if` statement. You use { } to tell the system which commands go with this `else if` clause.

- ▶ `system("espeak \"moving robot\"")`: This executes `espeak`, which should tell you that you are about to run your robot program.

- ▶ `system("/home/ubuntu/maestro_linux/robot.py")`: This is the program you will execute. In this case, your mobile platform will do whatever the program `robot.py` tells it to do.

After doing this, you will need to recompile the program, so type `make` and the executable `pocketsphinx_continuous` will be created. Run the program by typing `./pocketsphinx_continuous`. Disconnect the LAN cable, and the mobile platform will now take the "forward" voice command and execute your program.

Objective complete – mini debriefing

You should now have a complete mobile platform! When you execute your program, the mobile platform can now move around based on what you have programmed it to do.

Classified intel

You don't have to put all of your capabilities in one program. You can create several programs, each with a different function, and then connect each of the programs to their appropriate voice commands. Perhaps, one command moves your robot forward, a different backwards, another to turn right or left.

Mission accomplished

Congratulations! Your robot should now be able to move around in any way you'd like to program. You can even have the robot dance.

A challenge

You've now built a two-legged robot, and you can easily expand this to robots with even more legs. Here is an image of the mechanical structure of a four-legged robot that has eight DOF, which is fairly easy to create using many of the parts you have used to create your two-legged robot. This is my personal favorite because it doesn't fall over and break the electronics:

You'll need eight servos, and lots of battery. If you look on eBay, you can often find kits for sale for four-legged robots with twelve DOF, but again realize that the battery will need to be much bigger. For these kinds of applications, we often use **remote control** (**RC**) batteries. These are nice, as they are rechargeable, but make sure you either purchase one that is 5 to 6 volts, or include a way to regulate the voltage. Here is a picture of this kind of battery, available at most hobby stores:

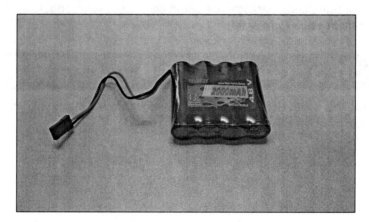

If you use this type of battery, don't forget a charger. The hobby store can help with choosing an appropriate match.

7

Avoiding Obstacles Using Sensors

In the previous two chapters, we covered wheeled and tracked movement and then legged movement. Now your robot can move around. But what if you want the robot to sense the outside world, so you don't run into things? In this project, you'll discover how to add some sensors to help us avoid barriers.

Mission briefing

We've covered creating robots using a wheeled/track base and robots that can move using legs. In this chapter, you will be introduced to some of the basics of sensors, particularly sensors that can help you avoid objects by alerting you to them.

Why is it awesome?

Your robot will take quite a beating if it continually runs into walls, or off the edge of a surface. Let's help your robot avoid these so that it looks intelligent.

Your objectives

In this chapter, you will:

- Connect the BeagleBone Black to a **USB sonar sensor** to detect the world around it
- Use a **servo** to change the position of your sensor so that a single sensor can view a large field, eliminating the need for additional sensors

Downloading the example code and colored images

You can download the example code files and colored images for this Packt book you have purchased from your account at `http://www.packtpub.com`. If you purchased this book elsewhere, you can visit `http://www.packtpub.com/support` and register to have the files e-mailed directly to you.

Mission checklist

In this chapter, you'll need some sensors. I am going to show you how to interface BeagleBone Black with a sonar sensor. There are other possibilities as well. IR sensors are also available that will detect distance from a target. However, they are sometimes difficult to interface using USB. The advantage of this particular sensor is that it already comes with a USB interface. Here is a image of the USB sonar sensor I like to use on my projects:

It is the **USB-ProxSonar-EZ**, and can be purchased directly at MatBotix or on Amazon. There are several models, each with a different distance specification; however, they all work in the same manner.

Also, you may want to detect distance in more than one direction. You have two choices. The first is simply to use a number of these sensors, one in each direction. But in the second section of this chapter, I'll show you how to use a servo to rotate the sensor. This way you can use a single sensor and just turn it to a new direction. To complete this section, you'll need a servo and a way to mount it to your project. Again, I like the Hitec series of servos, and this is ready-made for an HS-311 servo, which should look like this:

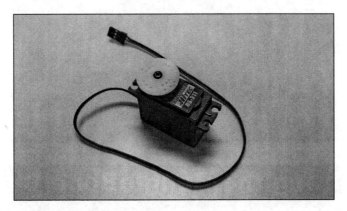

Here is a way to mount the sensor to a 90 degree angle bracket. I used one from a robot kit I purchased on eBay. It can connect to the servo like this:

However, if you want to get really fancy, you can purchase a pan-and-tilt assembly. These contain two servos and they allow you to rotate your sensor in both the vertical and horizontal axis. They are available from online stores such as www.robotshop.com. You can also construct a pan-and-tilt assembly out of components that you may have if you purchased a legged robot kit.

The finished product with servos looks like this:

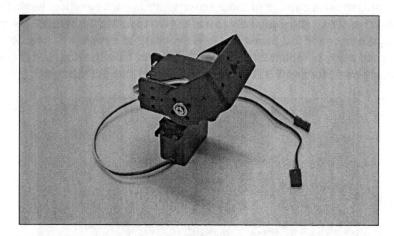

Connecting the BeagleBone Black to a USB sonar sensor

Now that you have a mobile platform and your robot can move, you will want to check if your robot could run into something. One of my favorite ways to do this is using a sonar sensor. First: a little tutorial on sonar sensors. This type of sensor uses ultrasonic sound to calculate the distance to an object. The sound wave travels out from the sensor, as illustrated here:

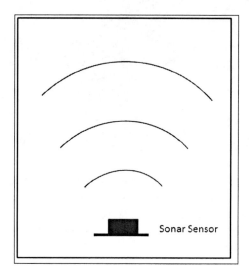

The device sends out a sound wave ten times a second. If an object is in the path of these waves, then the waves reflect off the object, sending waves that return to the sensor, as shown here:

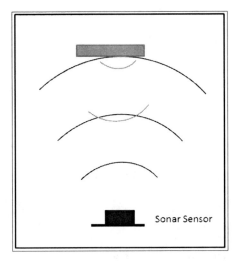

The sensor then measures any return. It uses the time difference between when the sound wave was sent out and when it returned, to measure the distance to the object.

Prepare for lift off

The first thing you'll want to do is connect the USB sonar sensor to your PC, just to make sure everything works well. Here are the steps:

1. First, download the terminal emulator SW from `http://www.maxbotix.com/articles/059.htm`, and select the Windows download. The page will look like the following screenshot:

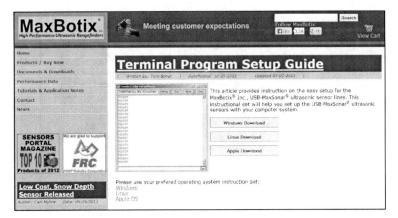

2. Unzip this file. Then plug the sensor into a USB port on your PC. Then open the terminal emulator file by selecting this file from the directory, as shown in the following screenshot:

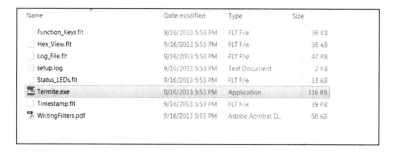

3. The following application window should pop up:

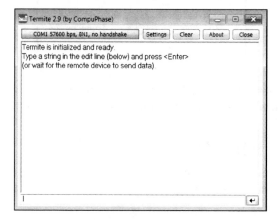

4. You'll need to change the setting to find the sensor, so select the **Settings** button, and you should see this screen:

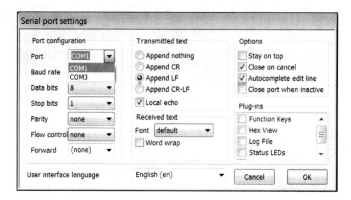

5. Select the **Port** menu and select the port that is connected to your sensor. Most often this will be the last one in the list. In my case, I selected **COM3**, clicked on **OK**, and this is what I saw on the main screen:

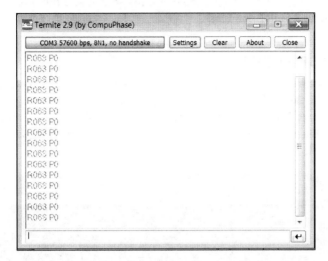

6. Notice the sensor readings. Now place an object in front of the sensor. You should now see something like this:

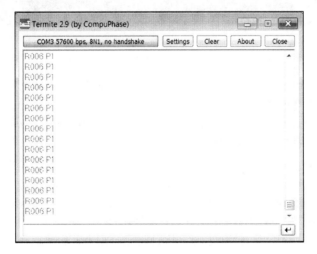

The readings have changed, specifically the value after **R**, and the value **P1**, indicating an object in front of the sensor. The **R** values indicates a range in mm, and **P1** indicates an object is in the range of the sensor. It will read **P0** if it thinks there is no object. You'll need to read these values into your program, and then you can avoid the object.

Now that you know the unit works, you'll want to mount the USB sensor on! your mobile platform. In this case, I am going to mount the USB sonar sensor on my quadruped robot.

Make sure you plug one end of the USB cable into the sensor and the other end into the USB hub connected to the BeagleBone Black.

Engage thrusters

With the HW all constructed and the sensor working, you can start communicating with your USB sensor using the BeagleBone Black. You are going to create a simple Python program that will read the value from the sensor. To do this—using **Emacs** as an editor—type `emacs sonar.py`. A new file will be created called `sonar.py`. Then type the code shown in the following screenshot:

```
ubuntu@ubuntu-armhf: ~/maestro_linux
File Edit Options Buffers Tools Python Help
#!/usr/bin/python
import serial

if __name__ =="__main__":
    ser=serial.Serial('/dev/ttyUSB0', 57600, timeout = 1)
    x = ser.read(100)
    print(x)
```

Let's go through the code to see what is happening.

- `#!/usr/bin/python`: This line simply makes this file available for us to execute from the command line.

- `import serial`: You also again import the serial library. This will allow us to interface with the USB sonar sensor.

- `if __name__ =="__main__":` The main part of the program is then defined.

- `Ser=serial.Serial('/dev/ttyUSB0', 57600, timeout = 1)`: This command sets up the serial port to use the `/dev/ttyUSB0` device, which is the sonar sensor, using a baud rate of 57600 and a timeout of 1.

- `x = ser.read(100)`: This command then reads the next hundred values from the USB port.

- `print(x)`: This command then prints out the value.

Once you have this file created, you can run the program and communicate with the device. Do this by typing `./sonar.py`, and the program will run. I have found that sometimes the device returns no data when run for the first time, so don't be surprised if you print out no values the first time you run your program. For the second time, you should receive a valid return string. Here is my result after running the program:

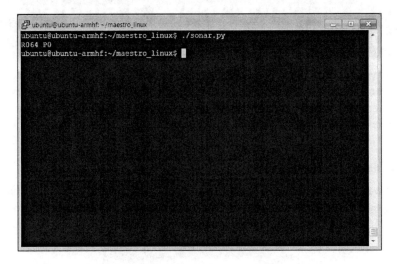

The sensor is returning `064`, which indicates a relative distance to a barrier in millimeters. If I place a good reflector just a few inches in front of the sensor and run the program, I will get the following result:

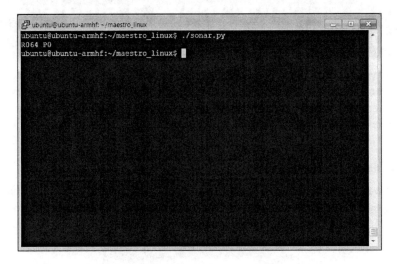

Objective complete – mini debriefing

Now the robot can sense its environment, so you can avoid bumping into walls and other barriers!

Using a servo to move a single sensor

You now have your sensor, but if you want to sense in more than just one direction, you could use several sensors, each mounted to a different side of the robot. However, there is a way to use servos to move your sensor, which allows you to use a single sensor to sense in several directions.

Prepare for lift off

The simplest way to avoid having to purchase and configure several sensors is to mount the sensor on a single servo, then use a servo bracket to connect this assembly to the platform. Using the sonar sensor, the assembly will look something like this:

Make sure you connect your servo to the servo controller; it can fit into any open connection. I am connecting mine to my quadruped that has eight servos to control, so I have connected mine to the eighth connection on the servo controller board. Here is an image:

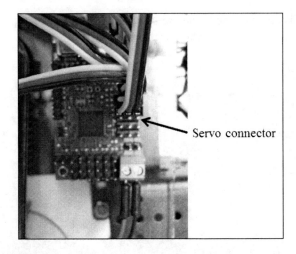

Servo connector

Engage thrusters

I'll assume you already have your sensor up and working and know how to read data. In this section, you will add the ability to move the sensor by communicating with the servo through the servo controller you configured in the last chapter.

For the program, you will begin with the `robot.py` program you created in *Chapter 6, Making the Unit Very Mobile - Controlling Legged Movement*, as you are going to need to access the servo controller. However, you may want to keep a copy of this program, just in case you want to use it later. First go to the directory that holds the `robot.py` program; in my case I placed it in the `maestro_linux` directory, so I would type `cd ./maestro_linux` from my `log-in` or `home` directory. Now let's create a copy of this program by typing `cp robot.py sense.py`.

You'll want to edit this program, so if you are using the Emacs editor, type `emacs sense.py`. The program you want to create will look like this:

```
ubuntu@ubuntu-armhf: ~/maestro_linux
File Edit Options Buffers Tools Python Help
#!/usr/bin/python
import serial
import time
class PololuMicroMaestro(object):
    def __init__(self, port= "/dev/ttyACM0"):
        self.ser = serial.Serial(port = port)
    def setAngle(self, channel, angle):
        minAngle = 0.0
        maxAngle = 180.0
        minTarget = 256.0
        maxTarget = 13120.0
        scaledValue = int((angle / ((maxAngle - minAngle) / (maxTarget - minTar\
get))) + minTarget)
        commandByte = chr(0x84)
        channelByte = chr(channel)
        lowTargetByte = chr(scaledValue & 0x7F)
        highTargetByte = chr((scaledValue >> 7) & 0x7F)
        command = commandByte + channelByte + lowTargetByte + highTargetByte
        self.ser.write(command)
        self.ser.flush()
    def close(self):
        self.ser.close()
if __name__ == "__main__":
    robot = PololuMicroMaestro()
    sensor=serial.Serial('/dev/ttyUSB0', 57600, timeout = 1)
    robot.setAngle(8,65)
    time.sleep(2.5)
    range = sensor.read(100)
    print(range)
    robot.setAngle(8,90)
    time.sleep(2.5)
    range = sensor.read(100)
    print(range)
    robot.setAngle(8,115)
    time.sleep(2.5)
    range = sensor.read(100)
    print(range)
-=--:**--F1  sense.py      All L36    (Python)---------------------------------
```

Let's walk through the code to see what it does. I will begin with the section that begins with `if __name__="__main__":`, as everything above this comes to us from the `robot.py` code and is covered in this previous chapter.

- ▸ The `robot=PololuMicroMaestro()` line initializes the servo motor controller and connects it to the proper USB port.

- ▸ `sensor=serial.Serial('/dev/ttyUSB0', 57600, timeout = 1)` opens a serial port that connects you to the USB sonar sensor at the `/dev/ttyUSB0` port and sets its parameters.

▸ You can now ask the servo to go to a specific position and then take a reading. In this case, I am doing this for the servo positions at 65 degrees, 90 degrees, and 115 degrees. At each of these locations, you ask for a range reading. Notice that you need to wait 2.5 seconds for the sensor to respond, based on the specifications of the manufacturer for the device to deliver a stable reading.

Objective complete – mini debriefing

That's it! Now you can sense in front of you and to either side. Here is an example of what might be displayed as a result of running the program:

Classified intel

If you are adding the sensor/servo combination to your wheeled vehicle, you'll need to add the servo motor controller as well. The motor controllers, the servo controllers, and the USB sonar or IR sensor can all coexist on the same BeagleBone Black. You'll need to merge the `dcmotor.py` and the `sense.py` programs so that you can access each individual capability.

Here is a listing of a possible program that does this:

```
ubuntu@ubuntu-armhf: ~/smc_linux

File Edit Options Buffers Tools Python Help

if __name__ =="__main__":
    motor1 = MotorControllerOne()
    motor2 = MotorControllerTwo()
    robot = PololuMicroMaestro()
    sensor=serial.Serial('/dev/ttyUSB0', 57600, timeout = 1)
    motor1.exitSafeStart()
    motor2.exitSafeStart()
    motor1.setSpeed(int(2000))
    motor2.setSpeed(int(-2000))
    time.sleep(.5)
    motor1.setSpeed(int(0))
    motor2.setSpeed(int(0))
    robot.setAngle(8,65)
    time.sleep(2.5)
    range = sensor.read(100)
    print(range)
    robot.setAngle(8,65)
    time.sleep(2.5)
    range = sensor.read(100)
    print(range)
    robot.setAngle(8,90)
    time.sleep(2.5)
    range = sensor.read(100)
    print(range)
    robot.setAngle(8,115)
    time.sleep(2.5)
    range = sensor.read(100)
    robot.setAngle(8,90)
    time.sleep(2.5)
    range = sensor.read(100)
    print(range)
    robot.setAngle(8,115)
    time.sleep(2.5)
    range = sensor.read(100)
    time.sleep(.5)
    motor1.close()
    motor2.close()

-=--:**--F1   sense.py      All L1      (Python)--------------------------------
```

In order to make this fit here, I have not included the `#include serial` and `time`, the `MotorControllerOne`, and `MotorControllerTwo` classes from the `dcmotor.py` file, and the `PololuMicroMaestro` class from the `robot.py` file. These would all need to be included, and then this main program would move the robot and then sense the environment around it. This would be a great starting point for your mobile robot code.

Mission accomplished

Congratulations! You can now detect and avoid walls and other barriers to your robot. You can also use these sensors to detect objects that you might want to find.

A challenge

One way to use these sensors is to use two sensors to actually "find" an object. This can help your robot actually find the position of specific obstacles. How this is accomplished is detailed on the maxbotix.com website at: `http://www.maxbotix.com/documents/ MaxBotix_Ultrasonic_Sensors_Find_Direction_and_Distance.pdf`. You have all the knowledge you need to add this type of capability to your robot.

8
Going Truly Mobile – Remote Control of Your Robot

Based on the previous projects, you now have mobile robots that can move around, accept commands, see, and even avoid obstacles. This project will teach you how to electronically communicate with your robot without any wires.

Mission briefing

You're mobile, but you still need to use your LAN cable if you want to communicate electronically with your project. In this project, you'll learn how to communicate without wires, yet still remain in control of your robot.

Why is it awesome?

As you send your device out into the world, you may still want to communicate with it electronically without connecting a cable. If you add this capability, you can change what your mobile robot is doing without any physical contact, and still remain in complete control of your project.

Your objectives

In this project you will be:

- ▸ Connecting the BeagleBone Black to a wireless USB keyboard
- ▸ Using the keyboard to control your project

Mission checklist

In this project, you'll learn how to connect to your device wirelessly. There are several ways to accomplish this. I'm going to show you how to do this with a standard USB wireless input device. This will let you control your robot without running cables.

However, first you will probably want to purchase a small LCD display for your BeagleBone Black. This will allow you to monitor what is going on with your project. In the initial projects you used a separate computer monitor for this. But an HDMI or DVI monitor is just too big and not really designed for mobile use. Fortunately, there are several inexpensive choices for small LCDs that connect right to the BeagleBone Black. The following image shows a device I have used for some of my projects:

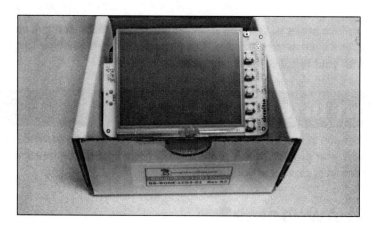

The LCD shown in the preceding image is from a company called CircuitCo. It's also available on Amazon and other online electronics stores, so you should be able to get it almost anywhere. There are several versions of LCD that are made for the BeagleBone Black. This one is 3.5 inches and has a display resolution of 320 x 240. There is another version that is 4.5 inches and has a display resolution of 480 x 270. The following image shows a larger LCD, made by 4D Systems:

The largest version is of 7 inches and has a display resolution of 800 x 480. I personally like the smaller versions for placement on my robots.

The board fits right on top of the BeagleBone Black. The following image shows an underside view of the LCD:

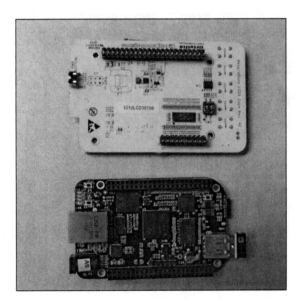

These pins are pushed right into the header connectors of the BeagleBone Black with the LCD on the top. The following image shows a side view of the BeagleBone Black and LCD:

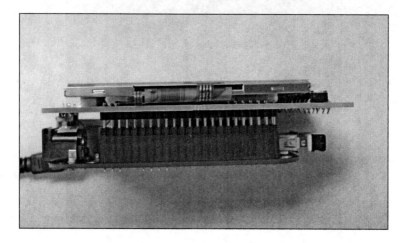

Now when you program your board, you can see the results right on the robotic platform. No extra programming is needed; the BeagleBone Ubuntu release will sense that the LCD screen is there and boot with the screen acting as a display. After the system boots, it will look similar to what is shown in the following image:

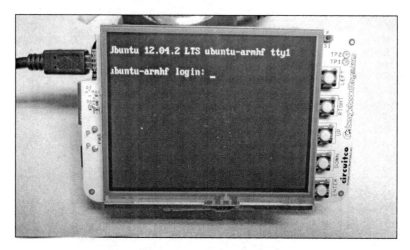

If you connect a USB keyboard to the BeagleBone Black and log in and then type `startx` at the prompt, you'll get your Xfce4 windowing system. It should look similar to what is shown in the following image:

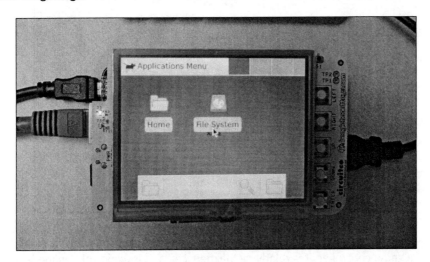

Now that you can display what is going on inside the BeagleBone Black, you need to choose which wireless input device to use.

A standard 2.4 GHz wireless keyboard is shown in the following image:

The preceding image displays a Logitech keyboard. Logitech generally makes a very reliable keyboard, and this connects well to the BeagleBone Black. This keyboard is available online through Amazon, and at most electronics or computer stores. You'll notice that this version has a built-in mouse.

Another choice is a small keyboard that looks more like a game controller. It will make your projects look amazing and will make it easier to control:

The preceding image shows a 2.4 GHz wireless keyboard by HausBell which is small, about the size of a game controller. It is relatively inexpensive, and is again sold online by Amazon.

There are several choices we could make for wireless technology in communicating with the BeagleBone Black. Bluetooth is quite popular, and works well, but comes with the added complexity of having to pair the device with the Bluetooth USB dongle and the system. The 2.4 GHz wireless technology comes with a keyboard and wireless USB receiver already paired so the device only works with the USB dongle that ships with the device. And the system automatically recognizes the device as long as the USB receiver is plugged into the USB port.

The frequency range utilized by 2.4 GHz wireless devices is the same as that of many 2.4 GHz wireless LAN devices, although they do not use the same modulation or protocol that is used by standard 2.4 GHz wireless LAN. Rather they use a proprietary modulation and protocol that is specific to the device and the company that manufactures the device.

While each device is different, most use the same overall approach, where they define a number of different channels, or small frequency ranges, inside the overall 2.4 GHz range. The keyboard communicates with the USB receiver on one of these frequencies. However, if either the keyboard or USB receiver sense that some other device is transmitting on that frequency, the devices will move to a different channel to try to avoid the interference.

The transmissions between the wireless keyboard and USB receiver are encrypted, so no device except the paired keyboard and USB receiver can understand the messages that are being sent between the two devices. The range of the keyboard/receiver pair is dependent upon the amount of power that both use for transmission, the higher the power the longer the range. Unfortunately, the higher the power the less time the batteries in the wireless device last. Most wireless keyboards are designed to work for up to 10 meters, or around 30 feet.

Connecting the BeagleBone Black to a wireless USB keyboard

You've been able to control your projects using a LAN connection, but you don't want to always have your projects tethered in this manner. In this section I'll show you how to connect via a wireless keyboard.

Prepare for lift off

Break out your USB keyboard. It should come with a USB dongle. Plug the USB dongle into the USB hub, and plug the hub into the BeagleBone Black USB port. If you are using a standard USB 2.4 GHz wireless keyboard, the entire system should look similar to what is shown in the following image:

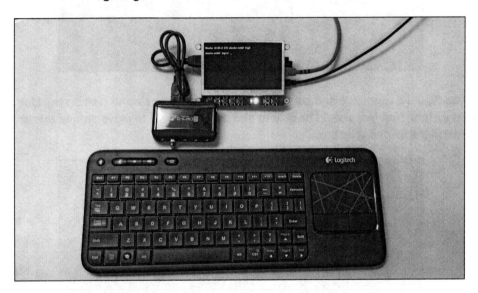

You can also use one of the 2.4 GHz wireless keyboards that look more like a gaming controller in the same manner.

Engage thrusters

Apply power to the USB hub and the BeagleBone Black. After some time, the unit should power on to display the log-in prompt. As you type in the username you should see the characters, similar to what is shown in the following image:

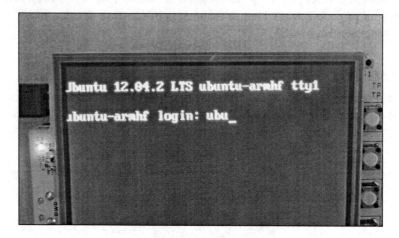

After you type the username and password, you can type `startx` and start up the Xfce window system. Now you should be able to also use the mouse to move around the screen, as shown in the following image:

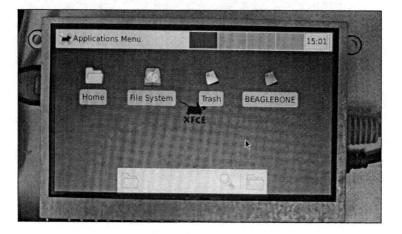

Objective complete – mini debriefing

You now have keyboard and mouse inputs. Next I'll show you how to accept the keystrokes into a program to control the robot.

Using the keyboard to control your project

Now the keyboard is connected, let's figure out how to accept commands on the BeagleBone Black.

Prepare for lift off

You can now enter commands wirelessly. The next step is to create a program that can take these commands and then have your project execute them. There are a couple of choices here and I'll give you examples of both. The first is simply to include the command interface in your program. Let's take the example of the program you wrote to move your wheeled robot, `robot.py`. If you like you can copy that program using `cp robot.py remote.py`. The following screenshot shows a listing of the current program in the area you want to change:

```
ubuntu@ubuntu-armhf: ~/smc_linux
File Edit Options Buffers Tools Python Help
if __name__ =="__main__":
    motor1 = MotorControllerOne()
    motor2 = MotorControllerTwo()
    print 'Ports created'
    motor1.exitSafeStart()
    motor2.exitSafeStart()
    print 'SafeStart'
    motor1.setSpeed(int(2000))
    motor2.setSpeed(int(-2000))
    print 'setSpeed'
    time.sleep(.5)
    motor1.setSpeed(int(0))
    motor2.setSpeed(int(0))
    time.sleep(.5)
    print 'at the end'
    motor1.close()
    motor2.close()
    print 'done'
-=--:----F1  dcmotor.py     Bot L57    (Python) -------------------------
```

Engage thrusters

In order to add user control, you need two new programming constructs: the `while` loop and the `if` statement. Let's add them to the program, and then I'll explain what they do. The following screenshot shows a listing of the area of the code you are going to change:

```
ubuntu@ubuntu-armhf: ~/smc_linux
File Edit Options Buffers Tools Python Help

if __name__ =="__main__":
    motor1 = MotorControllerOne()
    motor2 = MotorControllerTwo()
    motor1.exitSafeStart()
    motor2.exitSafeStart()
    var = 'n'
    while var != 'q':
        var = raw_input(">")
        if var == '<':
            motor1.setSpeed(int(2000))
            time.sleep(.5)
            motor1.setSpeed(int(0))
        if var == '>':
            motor2.setSpeed(int(-2000))
            time.sleep(.5)
            motor2.setSpeed(int(0))
        if var == 'f':
            motor1.setSpeed(int(2000))
            motor2.setSpeed(int(-2000))
            time.sleep(.5)
            motor1.setSpeed(int(0))
            motor2.setSpeed(int(0))
        if var == 'r':
            motor1.setSpeed(int(-2000))
            motor2.setSpeed(int(2000))
            time.sleep(.5)
            motor1.setSpeed(int(0))
            motor2.setSpeed(int(0))
    motor1.close()
    motor2.close()

-=--:----F1   remote.py      Bot L71     (Python)--------------------
```

You will edit your program by making the following changes. Add the following code just below the `if __name__=="__main__"` statement:

1. `var = 'n'`: This will define a variable named `var`, which will be of type character, which you will use in your program to get the input from the user.

2. `While var != 'q'`: This will put your program in a loop. This loop will repeat until you, or the user, enters the letter `q`.

3. `var - raw_input(">")`: This will get the character value from the user.

4. `If var == '<'`: This checks the value that you got from the user. If it is a < character, the robot will turn left by running the right DC motor for half a second. (You will need to determine how much time should be utilized to run the right DC motor for a left turn. The actual time value, `.5` in this case, may need to be higher or lower.)

5. Type the next few lines of code, which will send a speed command to the motor, wait for 0.5 seconds, and then send a command for the motor to stop.

6. `If var == '>'`: This checks the value that you got from the user. If it is a > character, the robot will turn right by running the left DC motor for half a second. (You will need to determine for how much time to run the left DC motor for a right turn. The actual time value, `.5` in this case, may need to be higher or lower.)

7. Type the next few lines of code, which will send a speed command to the motor, wait for 0.5 seconds, and then send a command for the motor to stop.

8. `If var == 'f'`: This checks the value that you got from the user. If it is an f character, the robot will run forward by running the right and left DC motors for half a second. (You will need to determine the speed to set each motor for a straight forward path.)

9. Type the next few lines of code, which will send a speed command to both the motors, wait for 0.5 seconds, and then send a command for both motors to stop.

10. `If var == 'r'`: This checks the value that you got from the user. If it is an r character, the robot will run backward by running the right and left DC motors for half a second. (You will need to determine the speed to set each motor for a straight backward path.)

11. Type the next few lines of code, which will send a speed command to both the motors, wait for 0.5 seconds, and then send a command for both motors to stop.

Once you have edited the program, save it and make it executable by typing chmod +x remote.py. Now you can run the program, but you must run it by typing the command using the wireless keyboard. If you are not yet logged in to the BeagleBone Black directly, make sure you can see the LCD screen and log in using the wireless keyboard. You can now disconnect the LAN cable; you will be able to communicate with the BeagleBone Black via the wireless keyboard. The system should look similar to what is shown in the following image:

Using the wireless keyboard and LCD screen, change the directory to the one that holds the remote.py program. In my case this file was in the /home/ubuntu/smc_linux directory, so I used the cd smc_linux command from my home directory. Now you can run the program by typing ./remote.py. The screen will display a prompt, and each time you type the appropriate command (<, >, f, or r) and press *Enter*, your robot should move. You need to be advised that the range of this technology is at best around 30 feet, so don't let your robot get too far away.

Objective complete – mini debriefing

Now you can move your robot around using the wireless keyboard! You can even take your robot outside. You don't need the LAN cable to run your programs because you can run them using the LCD display and keyboard.

Classified intel

There is one more change you can make so that you don't have to hit the *Enter* key after each input character is typed. In order to make this work, you'll need to add some import commands to your program and then add one function that can get a single character without pressing the *Enter* key. The following screenshot shows the first change you'll need to make:

```
ubuntu@ubuntu-armhf: ~/smc_linux
File Edit Options Buffers Tools Python Help
#!/usr/bin/python
import serial
import time
import tty
import sys
import termios
class MotorControllerOne(object):
    def __init__(self, port="/dev/ttyACM0"):
        self.ser=serial.Serial()
        self.ser.port= port
    def exitSafeStart(self):
        command = chr(0x83)
        self.ser.open()
        self.ser.write(command)
        self.ser.flush()
        self.ser.close()
    def setSpeed(self, speed):
-=--:----F1   remote.py      Top L1      (Python)--------------
For information about GNU Emacs and the GNU system, type C-h C-a.
```

You'll need to add `import tty`, `import sys`, and `import termios`. These are all the libraries that you'll need for your function to work. The following screenshot shows the function itself, and how you're going to use it:

```
ubuntu@ubuntu-armhf: ~/smc_linux
File Edit Options Buffers Tools Python Help
        def reset(self):
            self.ser.reset()
        def close(self):
            self.ser.close()
def getch():
    fd = sys.stdin.fileno()
    old_settings = termios.tcgetattr(fd)
    tty.setraw(sys.stdin.fileno())
    ch = sys.stdin.read(1)
    termios.tcsetattr(fd, termios.TCSADRAIN, old_settings)
    print '\ncodessed char is \'' + ch +'\'\n'
    return ch
if __name__=="__main__":
    motor1 = MotorControllerOne()
    motor2 = MotorControllerTwo()
    motor1.exitSafeStart()
    motor2.exitSafeStart()
    var = 'n'
    while var != 'q':
        var = getch()
        var = raw_input(">")
        if var == '<':
            motor1.setSpeed(int(2000))
            time.sleep(.5)
            motor1.setSpeed(int(0))
-=--:----F1   remote.py      56% L67      (Python)--------------
```

Copy the code from function `def getch():` into your program. I'm not going to explain it in detail, just know that it gets a single character without the need of pressing the *Enter* key after each keystroke. The `print` statement in the function is optional; I like to use it to map the different keys of my keyboard. Then instead of using `var = raw_input(">")` to get the character, use `var = getch()`.

 The program changes your terminal settings so when you run your program, you can no longer stop the program by typing *Ctrl + C*. You'll need to type q, and your terminal setting will be restored.

Mission accomplished

Congratulations! Now you can take your robot out into the big wide world. You can even use the LCD and keyboard to make changes to your program, although the smaller screen size makes this a bit difficult.

A challenge

Many users are comfortable using gaming keypads. There are several that come with a wireless connection. You could try connecting one of these to your robot if you want it to seem more like a real video game. In my system, I actually used the wireless keypad from HausBell and mapped the arrow keys at the top of the keyboard to tell my robot to go forward, backward, left, and right. I figured out which keystrokes these translated to by simply running my program and looking at the `print` statement in the program.

Using a GPS Receiver to Locate Your Robot

Based on the previous projects, you now have mobile robots that can move around, accept commands, see, and even avoid obstacles. This project will help you locate your robot while it moves, which can be useful for a robot that is fully autonomous.

Mission briefing

The robot is mobile, but let's not let it get lost. You're going to add a GPS receiver so that you can always know where you are.

Why is it awesome?

As you let your device free, you may not only want it to know where it is, but also to have a way of finding out if it has made it to the desired location. One of the coolest things to connect to the robot is a GPS location device. In this project, I'll show you how to connect a GPS receiver to your robot and then use it to move in the correct direction.

Your objectives

In this project we will cover the following:

- Connecting the BeagleBone Black to a GPS device
- Accessing the GPS programmatically and determining how to move to a location

Mission checklist

In this project, you'll need a GPS device. There are a lot of options, and they come with many different interfaces, but because we want to avoid using a soldering iron or other complex connection processes, we're going to choose one with a USB interface. Here is an image of a device I have used for some of my projects:

The model number is **ND-100S** from **GlobalSat**. It is small, inexpensive, and it supports Windows, Apple, and Linux, so our system should be able to interface with it. It is available on Amazon and other online electronics stores, so you should be able to get it almost anywhere. However, it does not have the sensitivity of some GPS devices. So if you will be using your robot in buildings or other locations that might stifle GPS signals, you should look for devices that are more sensitive.

Connecting the BeagleBone Black to a GPS device

Unpack your GPS device; it is time to get started.

Prepare for lift off

Before we get started, let me first give you a brief tutorial on GPS. GPS, which stands for **Global Positioning System**, is a system of satellites that transmits signals. GPS devices use these signals to calculate a position. There are a total of 24 satellites transmitting signals all around the earth at any given moment, but your device can only see the signal from a much smaller set of satellites.

Each of these satellites transmits a very accurate time signal that your device can receive and interpret. It receives the time signal from each of these satellites, and then based on the delay, the time it takes the signal to reach the device, it calculates the receiver's position based on a procedure called triangulation. The following two diagrams illustrate how the device uses the delay differences from three satellites to calculate its position:

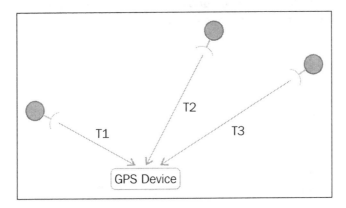

The GPS device is able to detect the three signals, and the time delays associated with receiving these signals. In the following diagram the device is at a different location, and the time delays associated with the three signals has changed:

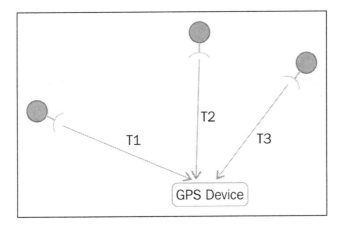

The time delays of the signals **T1**, **T2**, and **T3** can provide the GPS with an absolute position using a mathematical process called triangulation. Since the position of each satellite is known, the amount of time that the signal takes to reach the GPS device is also a measure of the distance between that satellite and the GPS device. To simplify, let's show an example in two dimensions. If the GPS device knows one distance to a satellite based on the amount of time delay, we could draw a circle around the satellite at that distance and know that our GPS device is on that sphere, as shown in this diagram:

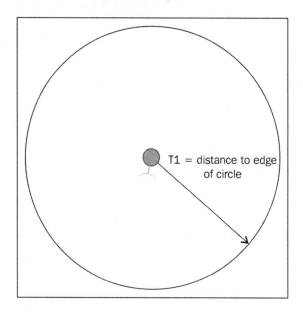

If we have two satellite signals and know the distance between the two, we can draw two circles as shown in the following diagram:

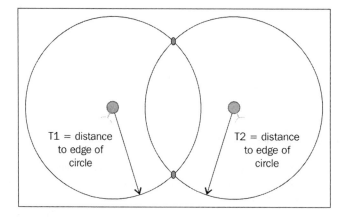

However, since we know that we can only be at points on the circle, we must be at one of the two points that are on both circles. Adding an additional satellite would eliminate one of these two points, providing us with an exact location. We need more satellites if we are going to do this in all the three dimensions.

Now, it is time to connect the device. As a first step, I suggest you connect the dongle to your PC. This will let you know the unit works and help you understand the device a little better. Then you'll connect it to the BeagleBone Black.

In order to install the system on your PC, insert the CD and run the setup program. You should see something like this:

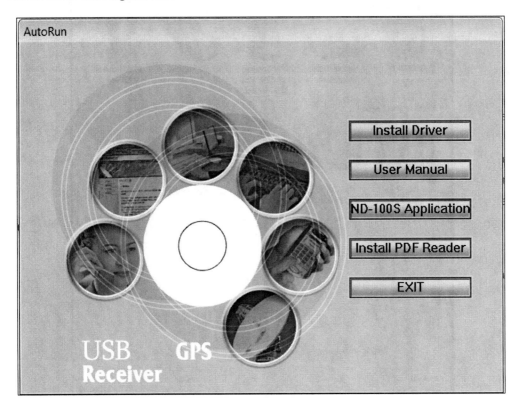

Click on both the **Install Driver** and **ND-100S Application** buttons and follow the default instruction procedures. When you have installed both the drivers and the application, you should be able to plug the GPS device into the USB port on your PC. A blue light at the end of the device should indicate that the device has been plugged in. The system will recognize the device, install the proper drivers, and you will have access to your device (this may take a few minutes). To ensure that the device has been installed, check your **Devices and Printers** start menu selection (if running Windows 7). You should see this:

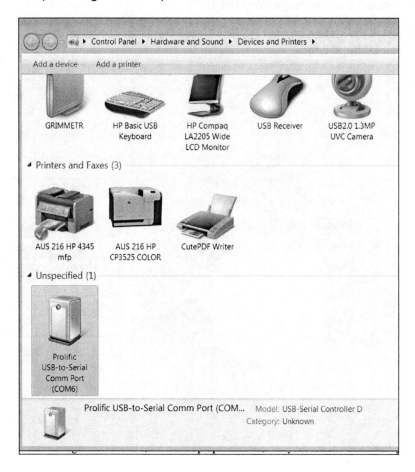

Once the device is installed, you can also run the application that comes on the CD-ROM. On startup it should look something like this:

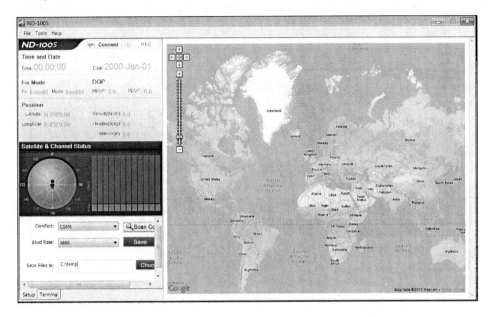

Now press the **Connect** selection button on the top-left of the screen. It should now look like this:

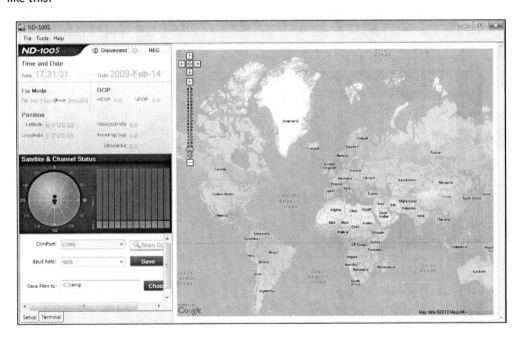

Unfortunately, if you are in a building or in a place where receiving information from the GPS satellites is difficult, the device may struggle to find its position. If you want to know that the system is working, even though it may struggle to find signals, select the **Terminal** tab selection in the lower-left corner. You should see something like this:

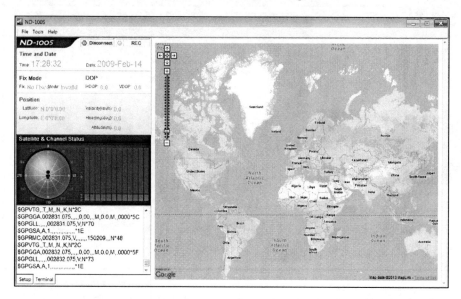

Notice that the lower-left window indicates the device is trying to find its location. Initially the unit in my office was unable to locate the satellites, not surprising for a building designed to restrict the transmission of signals into and out of the building. Following the same procedure on my laptop shows the following:

You'll notice that the blue LED on the end of the GPS is flashing. Now we have a position. If I select the **Terminal** tab, it shows the raw data coming back from the GPS:

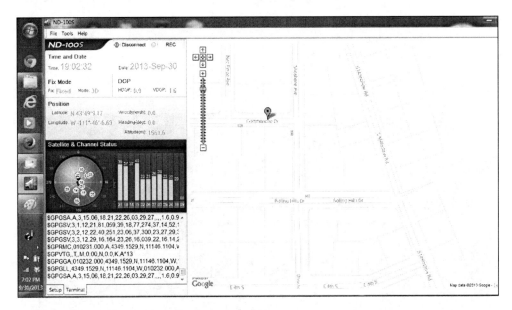

We'll use that raw data in our next section to plan our path to other positions. So, in an environment where GPS data is available, the unit is able to sync up and show your position. The next step will be to hook it to your BeagleBone Black robot.

Engage thrusters

First, connect the GPS unit by plugging it into one of the free USB ports on the USB hub. Once it is plugged in, and the unit is rebooted, type `lsusb` and you should see the following:

```
ubuntu@ubuntu-armhf:~$ lsusb
Bus 001 Device 002: ID 1a40:0101 Terminus Technology Inc. 4-Port HUB
Bus 001 Device 001: ID 1d6b:0002 Linux Foundation 2.0 root hub
Bus 001 Device 003: ID 1a40:0101 Terminus Technology Inc. 4-Port HUB
Bus 001 Device 004: ID 067b:2303 Prolific Technology, Inc. PL2303 Serial Port
Bus 001 Device 005: ID 1ffb:00a1
Bus 001 Device 006: ID 1ffb:00a1
Bus 001 Device 007: ID 046d:c52b Logitech, Inc. Unifying Receiver
ubuntu@ubuntu-armhf:~$
```

The device is shown as `Prolific Technology, Inc. PL2303 Serial Port`. Your device is now connected to your BeagleBone Black.

Now create a simple Python program that will read the value from the GPS device. If you are using Emacs as an editor, type `emacs measgps.py`. A new file will be created called `measgps.py`. Then type the following code:

```
ubuntu@ubuntu-armhf: ~/gps
File Edit Options Buffers Tools Python Help
#!/usr/bin/python
import serial

if __name__=="__main__":
    ser = serial.Serial('/dev/ttyUSB0', 4800, timeout = 1)
    x = ser.read(1200)
    print (x)




-=--:----F1   measgps.py      All L1      (Python)-----------------------
For information about GNU Emacs and the GNU system, type C-h C-a.
```

Let's go through the code to see what is happening.

1. `#!/usr/bin/python`: This line simply makes this file available for us to execute from the command line.

2. `import serial`: We import the serial library. This will allow us to interface with the USB GPS sensor.

3. `if __name__=="__main__"::` The main part of our program is then defined.

4. `ser = serial.Serial('/dev/ttyUSB0', 4800, timeout = 1)`: This command sets up the serial port to use the `/dev/ttyUSB0` device, which is our GPS sensor using a baud rate of `4800` and a timeout of `1`.

5. `x = ser.read(1200)`: This command then reads in a set of values from the USB port. In this case, we read `1200` bytes, which will include a fairly full set of our GPS data.

6. `print x`: This final command then prints out the value.

Once you have this file created, you can run the program and talk to the device. Do this by typing `python measgps.py` and the program will run. You should see something like the following:

```
ubuntu@ubuntu-armhf:~/gps$ python measgps.py
$GPGGA,160113.167,,,,,0,00,,,M,0.0,M,,0000*52
$GPGLL,,,,,160113.167,V,N*7E
$GPGSA,A,1,,,,,,,,,,,,,,,,*1E
$GPRMC,160113.167,V,,,,,,,011013,,,N*4B
$GPVTG,,T,,M,,N,,K,N*2C
$GPGGA,160114.170,,,,,0,00,,,M,0.0,M,,0000*53
$GPGLL,,,,,160114.170,V,N*7F
$GPGSA,A,1,,,,,,,,,,,,,,,,*1E
$GPRMC,160114.170,V,,,,,,,011013,,,N*4A
$GPVTG,,T,,M,,N,,K,N*2C
$GPGGA,160115.170,,,,,0,00,,,M,0.0,M,,0000*52
$GPGLL,,,,,160115.170,V,N*7E
$GPGSA,A,1,,,,,,,,,,,,,,,,*1E
$GPRMC,160115.170,V,,,,,,,011013,,,N*4B
$GPVTG,,T,,M,,N,,K,N*2C
$GPGGA,160116.167,,,,,0,00,,,M,0.0,M,,0000*57
$GPGLL,,,,,160116.167,V,N*7B
$GPGSA,A,1,,,,,,,,,,,,,,,,*1E
$GPGSV,1,1,00*79
$GPRMC,160116.167,V,,,,,,,011013,,,N*4E
$GPVTG,,T,,M,,N,,K,N*2C
$GPGGA,160117.167,,,,,0,00,,,M,0.0,M,,0000*56
$GPGLL,,,,,160117.167,V,N*7A
$GPGSA,A,1,,,,,,,,,,,,,,,,*1E
$GPRMC,160117.167,V,,,,,,,011013,,,N*4F
$GPVTG,,T,,M,,N,,K,N*2C
$GPGGA,160118.170,,,,,0,00,,,M,0.0,M,,0000*5F
$GPGLL,,,,,160118.170,V,N*73
$GPGSA,A,1,,,,,,,,,,,,,,,,*1E
$GPRMC,160118.170,V,,,,,,,011013,,,N*46
$GPVTG,,T,,M,,N,,K,N*2C
$GPGGA,160119.170,,,,,0,00,,,M,0.0,M,,0000*5E
$GPGLL,,,,,160119.170,V,N*72
$GPGSA,A,1,,,,,,,,,,,,,,,,*1E
$GPRMC,160119.170,V,,,,,,,011013,,,N*
ubuntu@ubuntu-armhf:~/gps$
```

The device is providing raw readings back to you, which is a good sign. Unfortunately there isn't much good data here, as the unit again is inside. How do we know this? Look at one of the lines that starts with $GPRMC. This line should tell us our current latitude and longitude values. The GPRS is reporting:

`$GPRMC,160119.170,V,,,,,,,011013,,,N*`

This line of data should take the following form, with each field separated by a comma:

0	1	2	3	4	5	6
$GPRMC	220516	A	5133.82	N	00042.24	W

7	8	9	10	11	12
173.8	231.8	130694	004.2	W	*7

And here is the table providing explanation of each of these fields:

Field	Value	Explanation
1	220516	Timestamp
2	A	Validity: A (OK), V (Invalid)
3	5133.82	Current latitude
4	N	North or south
5	00042.24	Current longitude
6	W	East or west
7	173.8	Speed in knots with which the device is moving
8	3	Course: The angle direction in which the device is moving
9	130694	Datestamp
10	0004.2	Magnetic variation: Variation from magnetic and true north
11	W	East or west
12	*70	Checksum

In our case, field 2 is reporting V, or that the unit cannot find enough satellites to get a position. Taking the unit outside, we can get something like this on our `measgps.py`:

```
ubuntu@ubuntu-armhf: ~/gps
ubuntu@ubYÆFÆ&ævÆV&æÆOcÒfæÆOcÆ¦VVkps.py
$GPGLL,4349.1422,N,11146.1055,W,020737.000,A,A*47
$GPGSA,A,3,22,21,15,18,,,,,,,,,5.4,2.7,4.7*3B
$GPRMC,020737.000,A,4349.1422,N,11146.1055,W,0.32,289.37,021013,,,A*77
$GPVTG,289.37,T,,M,0.32,N,0.6,K,A*0D
$GPGGA,020738.000,4349.1425,N,11146.1059,W,1,04,2.7,1521.1,M,-16.9,M,,0000*52
$GPGLL,4349.1425,N,11146.1059,W,020738.000,A,A*43
$GPGSA,A,3,22,21,15,18,,,,,,,,,5.4,2.7,4.7*3B
$GPRMC,020738.000,A,4349.1425,N,11146.1059,W,0.75,280.54,021013,,,A*7C
$GPVTG,280.54,T,,M,0.75,N,1.4,K,A*01
$GPGGA,020739.000,4349.1428,N,11146.1066,W,1,04,2.7,1521.0,M,-16.9,M,,0000*53
$GPGLL,4349.1428,N,11146.1066,W,020739.000,A,A*43
$GPGSA,A,3,22,21,15,18,,,,,,,,,5.4,2.7,4.7*3B
$GPRMC,020739.000,A,4349.1428,N,11146.1066,W,1.34,243.86,021013,,,A*78
$GPVTG,243.86,T,,M,1.34,N,2.5,K,A*07
$GPGGA,020740.000,4349.1426,N,11146.1064,W,1,04,2.7,1523.6,M,-16.9,M,,0000*55
$GPGLL,4349.1426,N,11146.1064,W,020740.000,A,A*41
$GPGSA,A,3,22,21,15,18,,,,,,,,,5.4,2.7,4.7*3B
$GPRMC,020740.000,A,4349.1426,N,11146.1064,W,1.82,214.11,021013,,,A*7B
$GPVTG,214.11,T,,M,1.82,N,3.4,K,A*06
$GPGGA,020741.000,4349.1422,N,11146.1068,W,6,00,50.0,1523.3,M,-16.9,M,,0000*6A
$GP
ubuntu@ubuntu-armhf:~/gps$ 
```

Notice that that `$GPRMC` line now reads this:

```
$GPRMC,020740.000,A,4349.1426,N,11146.1064,W,1.82,214.11,021013,,,A*
7B
```

Our values are now:

Field	Value	Explanation
1	020740.000	Timestamp
2	A	Validity: A (OK), V (Invalid)
3	4349.1426	Current latitude
4	N	North or south
5	11146.1064	Current longitude
6	W	East or west
7	1.82	Speed in knots with which the device is moving
8	214.11	Course: The angle direction in which the device is moving
9	021013	Datestamp

Field	Value	Explanation
10		Magnetic variation: Variation from magnetic and true north
11		East or west
12	*7B	Checksum

Objective complete – mini debriefing

Now you have some indication of where you are; however, it is in raw form that may not mean much. In the next section, we will figure out how to do something with these readings.

Accessing the GPS programmatically and determining how to move to a location

Now that you can access your GPS device, let's work on accessing the data programmatically.

Prepare for lift off

Your project should now have the GPS connected and have access to querying the data via the serial port. In this section, you will create a program to use this data to discover where you are, and then you can determine what to do with that information.

Engage thrusters

If you completed the last section, you should be able to receive the raw data from the GPS unit. Now you want to be able to take this data and do something with it, for example, find your current location and altitude and then decide if your target location is to the west, east, north, or south.

First, get the information out of the raw data. As noted earlier, our position and speed is in the $GPMRC output of our GPS. First, write a program to simply parse out a couple of pieces of info from that data. So open a new file (you can name it `location.py`) and edit it as follows:

```
ubuntu@ubuntu-armhf: ~/gps
File Edit Options Buffers Tools Python Help
#!/usr/bin/python
import serial

if __name__ == "__main__":
    ser = serial.Serial('/dev/ttyUSB0', 4800, timeout = 1)
    x = ser.read(500)
    pos1 = x.find("$GPRMC")
    pos2 = x.find("\n", pos1)
    loc = x[pos1:pos2]
    data = loc.split(',')
    if data[2] == 'V':
        print 'No location found'
    else:
        print "Latitude = " + data[3] + data[4]
        print "Longitude = " + data[5] + data[6]
        print "Speed = " + data[7]
        print "Course = " + data[8]




-=--:----F1  location.py     All L1      (Python)--------------------------------
For information about GNU Emacs and the GNU system, type C-h C-a.
```

Let's go through the code to see what is happening.

1. The `#!/usr/bin/python`: As always, this line simply makes this file available for you to execute from the command line.

2. `import serial`: You again import the serial library. This will allow you to interface with the USB GPS sensor.

3. `if __name__=="__main__":`: The main part of your program is then defined.

4. `ser = serial.Serial('/dev/ttyUSB0', 4800, timeout = 1)`: The first command sets up the serial port to use the `/dev/ttyUSB0` device, which is your GPS sensor using a baud rate of `4800` and a timeout of `1`.

5. `x = ser.read(500)`: This command then reads in a set of values from the USB port. In this case, you read `500` values, which will include a fairly full set of your GPS data.

6. `pos1 = x.find("$GPRMC")`: This will find the first occurrence of `$GPRMC` and set the value of `pos1` to that position. In this case, you want to isolate the `$GPRMC` response line.

7. `pos2 = x.find("\n", pos1)`: This will find the end of this line.

8. `loc = x[pos1:pos2]`: The variable `loc` will now hold the line with all the information you are interested.

9. `data = loc.split(',')`: This will take your comma separated line and break it into an array of values.

10. `if data[2] == 'V'`:: You now check to see if the data is valid. If not, the next line simply prints out that you did not find a valid location.

11. `else`:: If the data is valid, the next few lines print out the various pieces of data.

Here is a screenshot showing the results when my device was able to find its location:

Once you have the data, you can do some interesting things with it. For example, you might want to figure out the distance and direction to another waypoint. There is a piece of code at `http://code.activestate.com/recipes/577594-GPS-distance-and-bearing-between-two-GPS-points/` that you can use to find the distance and bearing to other waypoints based on your current location. You can easily add this code to your `location.py` file to update your robot on the distance and bearing to other waypoints.

Objective complete – mini debriefing

Now your robot knows where it is and the direction it needs to go to get to other locations!

Classified intel

There is another way to configure your GPS device that may make it a bit easier to access the data from other programs. It is a set of functionality held in the `gpsd` library. To install this capability, type `sudo apt-get install gpsd gpsd-clients`, and this will install the gpsd SW. This SW works by starting a background program (called a `daemon`) that communicates with your GPS device. We can then just query the program to get the data. To make sure this works, type `cgps`, and a program that was installed with the `gpsd` library will open, and you should see this:

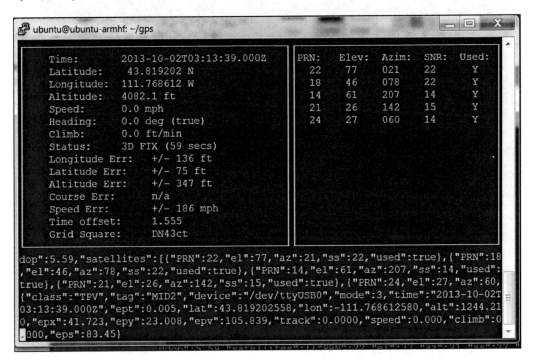

This displays both the formatted data and some of the raw data that is coming from the GPS sensor. We can also access this information from a program. To do this, edit a new file called `gpsd.py` as shown in the following screenshot:

```
ubuntu@ubuntu-armhf: ~/gps
File Edit Options Buffers Tools Python Help
#!/usr/bin/python
import gps

session = gps.gps("localhost", "2947")
session.stream(gps.WATCH_ENABLE | gps.WATCH_NEWSTYLE)
while True:
    report = session.next()
    if report['class'] == 'TPV':
        if hasattr(report, 'time'):
            print report.time




-=--:----F1  gpsd.py          All L1        (Python)-----------------------
For information about GNU Emacs and the GNU system, type C-h C-a.
```

Here are the details of your code:

1. `#!/usr/bin/python`: As always, the first line simply makes this file available for you to execute from the command line.

2. `import gps`: In this case you import the GPS library. This will allow you to access the gpsd functionality.

3. `session = gps.gps("localhost", "2947")`: This opens a communication path between the gpsd functionality and our program. This opens port `2947`, assigned to the gpsd functionality, on the local host.

4. `session.stream(GPS.WATCH_ENABLE | GPS.WATCH_NEWSTYLE)`: Tells the system to look for new GPS data as it becomes available.

5. `while True:`: This simply loops and processes information until you ask the system to stop (by typing *Ctrl + C*).

6. `report = session.next()`: When a report is ready it goes into the variable `report`.

7. `if report['class'] == 'TPV':`: Checks to see if the report will give you the type of report that you need.

8. `if hasattr(report, 'time'):`: Makes sure that `report` holds time data.

9. `print report.time`: Prints out the time data. I use this in my example because it is always returned, even if the GPS is not able to see enough satellites to return position data. To see other possible attributes, see `www.catb.org/gpsd/gpsd_json.html` for details.

Once you have created the program, you can run it by typing `python gpsd.py`. This is a possible output of running the program:

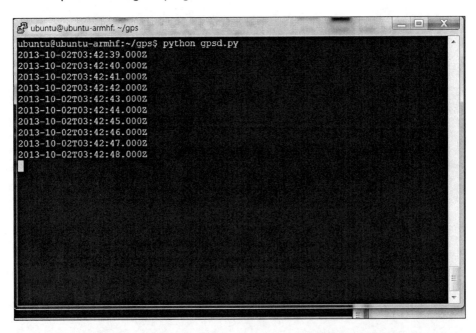

Mission accomplished

Congratulations! Your robot can now get around without getting lost. You can use the info to plan routes to different waypoints and track where your robot has been.

A challenge

One of the ways to display positional information is to use a graphical display including a map of your current position. There are several map applications that can interface with your GPS to indicate your location on a map. Here is an excellent tutorial on this: `https://www.sparkfun.com/tutorials/403`. You won't need to execute the HW configuration part of the tutorial, but will be able to start with the section **Read a GPS and plot position with Python**.

10

System Dynamics

Through the previous chapters we've spent time talking a lot about individual functionality that we can add to our robotic projects. In this chapter, we'll talk about how to integrate these different parts into a single system.

Mission briefing

We've spent lot of time on individual functionality, and your robotic projects now have lots of functionality that we can add to our projects. This chapter will bring all of these parts together into a framework that allows the different parts to work together.

Why is it awesome?

You don't want the robot to just walk, talk, or see. You want it to do all of these in a coordinated package. In this chapter, you'll learn how to programmatically connect all of these individual capabilities and make your projects seem intelligent.

Your objectives

In this chapter, we will:

- ▶ Create a general control structure so that different capabilities can work together through system calls

Mission checklist

Finally we're done purchasing the HW. In this chapter, we'll be adding functionality via SW. You'll need ample storage space for an array of new SW. First, let's check how much space you have on your memory card. You should install **discus**: a program that will let you see how much disk space you have used and how much is available for new programs.

To do this, type `sudo apt-get install discus`. This will install the discus application. Now run the program by typing `discus`. You should get something similar to the following screenshot:

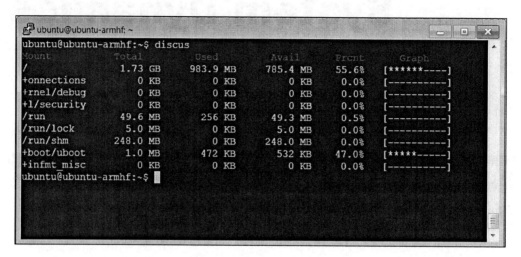

Notice that at this point my default drive only has 1.73 GB. The card I have is an 8 GB card, so how do I access the rest? I've already laid out the steps in *Chapter 4, Allowing the BeagleBone Black to See*, in the section on installing OpenCV. After I followed that set of instructions, I now see this:

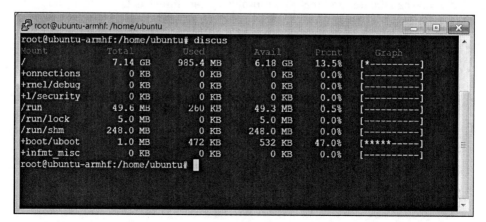

I now have a little over 7 GB of space to add capability. You can also use the command `df -h` to see this same information.

Creating a general control structure so capabilities can communicate

Now that you have a mobile robot, you want to coordinate all of its different abilities. Let's start with the simplest approach: using a single control program that can call other programs and enable all the capabilities.

Prepare for lift off

You've already done this once. In *Chapter 3, Providing Speech Input and Output*, you edited the `continuous.c` code to allow it to call other programs to execute functionality. Here is the code that we used, found in the `/home/ubuntu/pocketsphinx-0.8/programs/src/` directory.

```
File Edit Options Buffers Tools C Help
        fflush(stdout);
        /* Finish decoding, obtain and print result */
        ps_end_utt(ps);
        hyp = ps_get_hyp(ps, NULL, &uttid);
        printf("%s: %s\n", uttid, hyp);
        fflush(stdout);
        /* Exit if the first word spoken was GOODBYE */
        if (hyp) {
            sscanf(hyp, "%s", word);
            if (strcmp(hyp, "GOODBYE") == 0)
            {
                system("espeak \"good bye\"");
                break;
            }
            else if (strcmp(hyp, "HELLO") == 0)
            {
                system("espeak \"hello\"");
            }
        }

        /* Resume A/D recording for next utterance */
        if (ad_start_rec(ad) < 0)
            E_FATAL("Failed to start recording\n");
    }
-=-:----F1  continuous.c   79% L319   (C/l Abbrev)--------------
```

The functionality that is important to us is the `system("espeak \"good bye"\"");"\"");` line of code. When you use the `system` function call, the program actually calls a different program, in this case the `espeak` program, and passes it to the `good bye` parameter so that the words `good` and `bye` come out of the speaker.

Here is another example, this time from *Chapter 5, Making the Unit Mobile – Controlling Wheeled Movement*, when we wanted to command our robot to move:

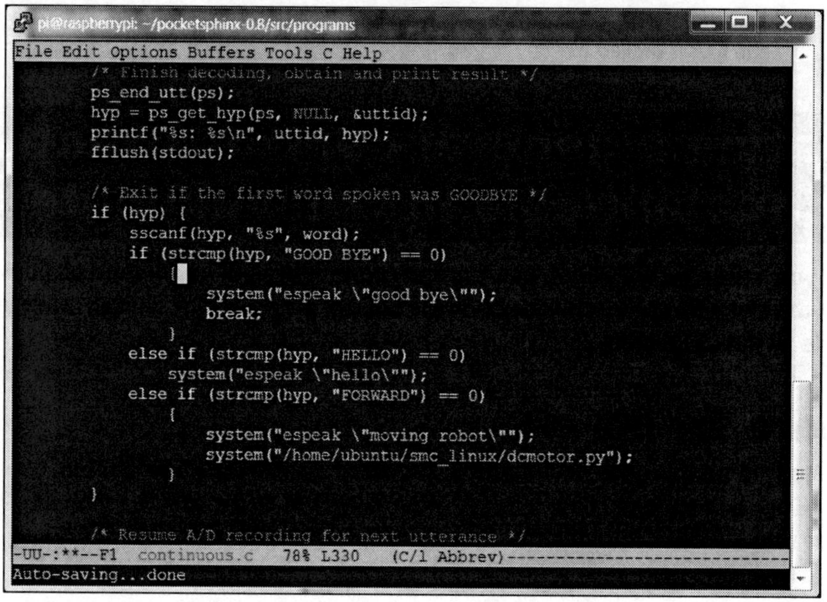

In this case, if you say `forward` to your robot, it will execute two programs. The first program you call is the `espeak` program with the parameter `moving robot`, and these words should then come out of the speaker on the robot. The second program is the `dcmotor.py` program, which should include the commands to move the robot forward.

I am now going to include an example in Python; it is my preferred language. I am going to use my wheeled robot:

It has a camera and is also able to communicate via a speaker. I control it via my wireless keyboard. I want to add the functionality to follow a colored ball, turn as the ball goes right or left, and tell me when it is turning.

You also need to make sure all of your devices are available to your programs. To do this, you'll need to make sure your USB camera as well as the two DC motor controllers are connected. To connect the camera, follow the steps given in *Chapter 4, Allowing the BeagleBone Black to See*, in the section *Connecting the USB Camera to the BeagleBone Black and viewing the Images*. It works best to connect the USB camera first before connecting any other USB devices.

After the camera is up and running, check that both motor controllers are available to the system. To do this, type `cd /home/ubuntu/scm_linux` and then type `.SmcCmd -list`. You should see something like this:

```
ubuntu@ubuntu-armhf: ~/smc_linux
ubuntu@ubuntu-armhf:~/smc_linux$ ./SmcCmd --list
2 Simple Motor Controllers found:
18v7 #51FF-7306-4980-4956-1527-1287
18v7 #51FF-7406-4980-4956-5842-1287
ubuntu@ubuntu-armhf:~/smc_linux$
```

Both DC motor controllers are available. The numbers are the serial numbers of each individual motor; you can use them to send commands to just one motor controller. For example, you may need to send the `resume` command to the motors. To do this type as shown in the following screenshot:

```
ubuntu@ubuntu-armhf: ~/smc_linux
ubuntu@ubuntu-armhf:~/smc_linux$ ./SmcCmd --list
2 Simple Motor Controllers found:
18v7 #51FF-7306-4980-4956-1527-1287
18v7 #51FF-7406-4980-4956-5842-1287
ubuntu@ubuntu-armhf:~/smc_linux$ ./SmcCmd -d 51FF-7406-4980-4956-5842-1287 --resume
ubuntu@ubuntu-armhf:~/smc_linux$ ./SmcCmd -d 51FF-7306-4980-4956-1527-1287 --resume
ubuntu@ubuntu-armhf:~/smc_linux$
```

I am going to involve three different programs. First, I am going to create a program that will find out if the ball is to the right or left. This is going to be my main control program. I am also going to create a program that moves my robot approximately 45 degrees to the right-hand side and another that moves my robot 45 degrees to the left-hand side. I am going to keep these very simple, and you might be tempted to just put them all in the same source file. But as the complexity of each of this program grows, it will make more sense for them to be separate, so this is a good starting point for your robotic code. Also, if you want to use the code in another project, or want to share the code, this sort of separation helps.

Engage thrusters

You are going to create three programs for this project. In order to keep this organized, I created a new directory in my home directory by typing `mkdir robot` in my home directory. I will now put all my files in this directory.

The next step is to create two files that can move your robot: one to the left-hand side, the other to the right-hand side. To do this, you will create two copies of the `dcmotor.py` code you created in *Chapter 5, Making the Unit Mobile – Controlling Wheeled Movement*, in your `robot` directory. If you have created that file in your `home` directory, type `cp dcmotor.py ./robot/move_left.py  cp dcmotor.py ./robot/move_right.py`. Now you'll edit those, changing two numbers in the `if __name__=="__main__":` command. Here is the edit to the `move_left.py` file:

```
ubuntu@ubuntu-armhf: ~/robot
File Edit Options Buffers Tools Python Help
        def reset(self):
            self.ser.reset()
        def close(self):
            self.ser.close()

if __name__ =="__main__":
    motor1 = MotorControllerOne()
    motor2 = MotorControllerTwo()
    motor1.exitSafeStart()
    motor2.exitSafeStart()
    motor1.setSpeed(int(2000))
    motor2.setSpeed(int(2000))
    time.sleep(.1)
    motor1.setSpeed(int(0))
    motor2.setSpeed(int(0))
    time.sleep(.1)
    motor1.close()
    motor2.close()

-=--:----F1  move_left.py    Bot L65    (Python)-------------
```

I'm not going to explain the details of the code; it is already detailed in *Chapter 5, Making the Unit Mobile – Controlling Wheeled Movement*. The numbers that I changed were the `motorx.setSpeed` numbers; using `2000` for both motors turns my robot to the left-hand side. Additionally, I changed the `time.sleep` numbers to `.1`, so the robot will respond more quickly. The `.1` will delay the execution of the program by one-tenth of a second. You will also need to edit `moveright.py` similarly:

```
ubuntu@ubuntu-armhf: ~/robot
File Edit Options Buffers Tools Python Help
        def reset(self):
            self.ser.reset()
        def close(self):
            self.ser.close()

if __name__ =="__main__":
    motor1 = MotorControllerOne()
    motor2 = MotorControllerTwo()
    motor1.exitSafeStart()
    motor2.exitSafeStart()
    motor1.setSpeed(int(-2000))
    motor2.setSpeed(int(-2000))
    time.sleep(.1)
    motor1.setSpeed(int(0))
    motor2.setSpeed(int(0))
    time.sleep(.1)
    motor1.close()
    motor2.close()

-=--:----F1  move_right.py    Bot L62    (Python)-------------
```

This time the `setSpeed` numbers are both -2000, turning my robot to the right-hand side.

The final step is to create our main control program. Let's call it `follow.py`. Open this file with your editor; if using Emacs type `emacs follow.py`:

```
ubuntu@ubuntu-armhf: ~/robot
File Edit Options Buffers Tools Python Help
#!/usr/bin/python
import cv2
import numpy
from subprocess import call

cap = cv2.VideoCapture(0)

while True:
    ret,img = cap.read()
    img = cv2.blur(img,(3,3))
    hsv = cv2.cvtColor(img,cv2.COLOR_BGR2HSV)
    threshold = cv2.inRange(hsv,numpy.array((0, 155, 0)), numpy.array((255, 255\
, 255)))
    contours, num = cv2.findContours(threshold,cv2.RETR_LIST,cv2.CHAIN_APPROX_S\
IMPLE)
    max_area = 0
    cx = 0
    cy = 0
    for cnt in contours:
        area = cv2.contourArea(cnt)
        if area > max_area:
            max_area = area
            max_cnt = cnt
    if max_area != 0:
        M = cv2.moments(max_cnt)
        cx,cy = int(M['m10']/M['m00']), int(M['m01']/M['m00'])
        cv2.circle(img,(cx,cy),5,255,-1)
    cv2.imshow("Ball Tracker", img)
    if cx > 280:
        call(["./move_right.py"])
    if cx < 20 and cx > 0:
        call(["./move_left.py"])
    if cv2.waitKey(10) == 27:
        break
-=--:----F1  follow.py      Top L1      (Python)----------------------
```

Let's look at this code:

1. `#!/usr/bin/python`: The first line allows your program to be run outside the Python environment. You'll use that later when you want to execute your code using `autostart` or using voice commands.

2. `import cv2`: The next line imports the OpenCV library. You need this to process the images.

3. `import numpy`: The next line imports the numpy library. This allows Python to work with the special arrays associated with OpenCV.

4. `from subprocess import call`: This library will allow you to call other programs from within your program.

5. `cap = cv2.VideoCapture(0)`: This line associates our program with the webcam.

6. `while True:`: Keep doing the loop; you'll only break if you press the *Esc* key in the image window.

7. `ret, img = cap.read()`: This line captures an image and moves it into the `img` array.

8. `img = cv2.blur(img, (3,3))`: This line smooths the images, getting rid of some of the random noise normally associated with images.

9. `hsv = cv2.cvtColor(img, cv2.COLOR_BGR2HSV)`: This OpenCV function converts the image file to the type you need to process it in a different color space.

10. `threshold = cv2.inRange(hsv, numpy.array((0, 155, 0)), numpy.array((255, 255, 255)))`: This creates a new image matrix, only allowing colors in a specific range. The `(0,155,0)` to `(255,255,255)` values let only green objects (the middle value has to be greater than `155`) through to the threshold image.

11. `contours, num = cv2.findContours(threshold, cv2.RETR_LIST, cv2.CHAIN_APPROX_SIMPLE)`: This finds the contours in the black and white image. These are places where there is a set of the same colors.

12. `max_area = 0, cx = 0, cy = 0`: These are all simple initializers.

13. `for cnt in contours:`: This section finds the biggest blob of colors; hopefully this will be the ball.

14. `if max_area != 0:`: If no set is found, we don't want to try to move to the proper `cx`, and `cy` pairs.

15. `M = cv2.moments(max_cnt)`: Find the moment (shape) associated with the biggest set of color.

16. `cx, cy = int(M['m10']/M['m00']), int(M['m01']/M['m00'])`: Find the center of the biggest set of color.

17. `cv2.circle(img, (cx, cy), 5, 255, -1)`: Draw a small blue circle on the image at the center of the biggest set of color.

18. `cv2.imshow("("Ball Tracker", ", img)`: Show the image on the screen.

19. `if cx > 280:`
    `call(["./(["./move_right.py"])"])`: If the x value of the biggest set of color is greater than `280`, then call the `move_right.py` program in this directory. This will move the robot to the right-hand side.

20. `if cx < 20 and cx > 0:`
    `call(["./(["./move_left.py"])"])`: If the x value of the biggest set of color is less than `20`, but greater than `0`, then call the `move_left.py` program in this directory. This will move the robot to the left-hand side.

21. `if cv2.waitKey(10) == 27:`: Stop the entire program if the *Esc* key is pressed while in the display window.

Objective complete – mini debriefing

Now you can run the program by typing `sudo ./follow.py`. The following window should be displayed:

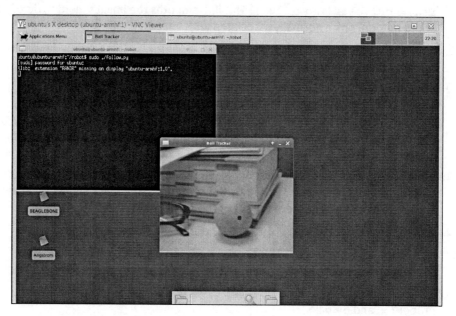

The blue dot indicates that the program is following the green ball. As the green ball is moved towards the left edge, the robot should also rotate slightly left. As the green ball is moved toward the right edge, the robot should also rotate slightly right.

Classified intel

You can change the color that you are looking for by changing the line `threshold = cv2.inRange(hsv,numpy.array((0, 155, 0)), numpy.array((255, 255,255)))`.

Two other color possibilities are:

▸ **Yellow:** `threshold = cv2.inRange(hsv,numpy.array((20, 100, 100)), numpy.array((30, 255,255)))`

▸ **Blue:** `threshold = cv2.inRange(hsv,numpy.array((100, 100, 100)), numpy.array((120, 255,255)))`

With OpenCV it is also possible to do motion detection. There are a couple of good tutorials on how to do this with OpenCV. One simple example is at `http://www.steinm.com/blog/motion-detection-webcam-python-opencv-differential-images/`. Another example, a bit more complex but more elegant, is at `http://stackoverflow.com/questions/3374828/how-do-i-track-motion-using-opencv-in-python`.

When using motion detection, if you roll the ball across the screen, you should see the following output on the webcam (using the code from the second tutorial):

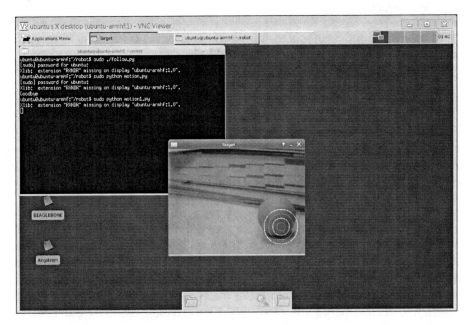

You can then use it to move the robot to follow the motion.

Mission accomplished

Now you can coordinate complex functionality for your robot. Your robot can walk, talk, see, hear, and even sense its environment, all at the same time.

A challenge

As you can see, communicating between different aspects of our project can be challenging. You are probably used to using an operating system that provides you with much of the basic functionality that you need for your computer use. In this section I'm going to introduce you to a special operating system that is designed specifically for use with robotics projects, the **Robot Operating System (ROS)**. This operating system sits on top of Linux and provides some interesting functionality.

ROS is fortunately free and open source. It is a very complex set of functionalities, but if you spend some time learning it, you can start using some of the most comprehensive functionality being developed in robotics research today.

To install ROS for the BeagleBone Black, go to `http://wiki.ros.org/groovy/Installation/UbuntuARM`. This gives you a step-by-step set of instructions to download and install the ROS onto your BeagleBone Black. Then also select Ubuntu on ARM, which is the architecture of your BeagleBone Black, and then select the appropriate version for the version of Ubuntu you are running. If you are using Ubuntu 12.04, for example, you'll want to select the 12.04 Precise armhf directions. Armhf is the architecture we are using, the ARM processor and hard float, which is what our processor supports.

Once installed, you can go through the tutorials; they will introduce you to the features of ROS and how to use it in our robotics projects. There are some limitations to using ROS on the BeagleBone Black; some of the nice graphical tools for monitoring and controlling the system are not available. However, it does provide a systematic way of configuring and communicating between multiple features of your robot running in different programs. It even comes with some programs that implement some interesting vision and motor control capabilities.

11

By Land, Sea, and Air

You've built robots that can navigate on land; now let's look at some possibilities for utilizing the tools to build some robots that dazzle the imagination.

Mission briefing

We've built robots that can navigate on land; now let's look at the possibilities for building robots that can navigate in the air or on the water. By now I hope you are comfortable accessing the USB control channels and talking with servo controllers and other devices that can communicate over USB. Instead of leading you through each step, in this chapter I'm going to point you in the right direction and allow you to explore a bit. I'll try to give you some examples using some of the projects that are going on around the Internet. I hope you are now ready to explore a bit on your own, for these projects can be quite complex and I'm not going to lead you through each step.

Why is it awesome?

You don't want to limit your robotic possibilities to just walking or rolling. You'll want your robot to fly, or sail, or swim. In this chapter, you'll see how you can use the capabilities you have already mastered in projects that defy gravity, explore the open sea, or navigate below the open sea.

Your objectives

In this chapter we will be:

- ▶ Using the BeagleBone Black in sailing robots
- ▶ Using the BeagleBone Black in flying robots
- ▶ Using the BeagleBone Black in submarine robots

Mission checklist

We need to add to our robotics HW in order to complete these projects. Since the HW is different for each of these projects, I'll introduce the HW in each individual section.

Using the BeagleBone Black in sailing robots

Now that you've created platforms that can move on land, let's turn to a completely different type of mobile platform—one that can sail. In this section, you'll discover how to use the BeagleBone Black to control your sail boat.

Prepare for lift off

Fortunately, sailing on the water is about as simple as walking on land. First, however, you need a sailing platform. The following image shows an RC sailing platform that can be modified to accept control from the BeagleBone Black:

In fact, many RC controller boats can be modified to add the BeagleBone Black. All you need is space to put the processor, the battery, and any additional control circuitry. In this case, the sailing platform has two controls: a rudder that is controlled by a servo and a second servo that controls the position of the sail. These are shown in the following image:

To automate the control of the sail boat, you'll need your BeagleBone Black, a battery, and a servo controller. The servo controller I would advise for this project is the little brother to the servo controller you used in *Chapter 5, Making the Unit Mobile – Controlling Wheeled Movement*. It is a six-servo controller made by Pololu, available at http://www.pololu.com, and it looks similar to what is shown in the following image:

The advantage is that this servo controller is much smaller and fits in limited space. The only challenge is creating a power connection to the device. Fortunately there is a cable that you can purchase that makes these power connections available from a standard cable. The cable you want is a USB to TTL serial/RS232 adapter cable. Make sure that the TTL end of the cable has individual female connectors. You can get this cable at `http://www.amazon.com`, and also at `http://www.adafruit.com`. An image of the cable is shown as follows:

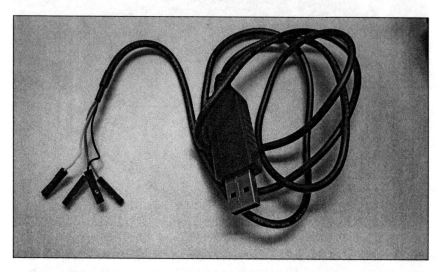

The red and black wires will be power. Now this can be connected to the servo controller, as shown in the following image:

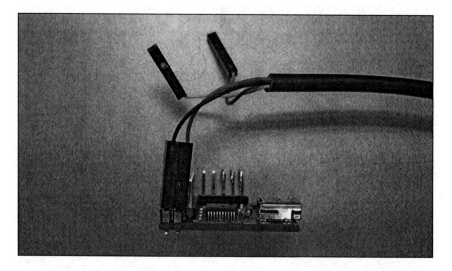

Engage thrusters

Once you have assembled your sail boat, you will need to first hook up the servo controller to the servos on the boat. You should try to control the servos before installing all the electronics inside the boat, similar to what is shown in the following image:

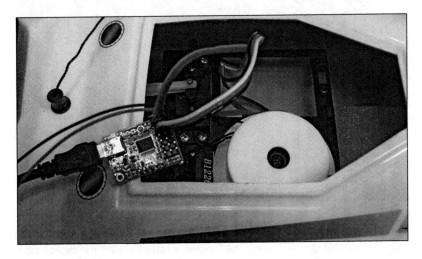

Just as in *Chapter 5, Making the Unit Mobile – Controlling Wheeled Movement*, you can use the MaestroController SW to control the servo controller from your PC. When you are ready to hook it up to the BeagleBone Black, you can start with the same Python program you used in *Chapter 5, Making the Unit Mobile – Controlling Wheeled Movement*. You will probably want to control the system without a wired connection, so you can use the principles that you learned in *Chapter 7, Avoiding Obstacles Using Sensors*.

It may be a bit challenging if you are using the standard 2.4 GHz keyboard, or smaller 2.4 GHz controller. You can add a bit more distance by connecting via a more powerful 2 way communications process. One possible solution is wireless LAN, unfortunately most lakes or ponds won't have an open wireless network available. You could also set up your own adhoc wireless network using a router connected to a laptop. Or many cell phones have the ability to set up a wireless hot spot, which can create a wireless network so that you can communicate remotely with your sailboat.

Another possible solution is to use ZigBee wireless devices to connect your sail boat to a computer. An image of a ZigBee device, called the XBee, is shown as follows:

You'll need two of them and a USB shield for each. You can get these at a number of places, including `http://www.adafruit.com`. The following image shows the device on the shield.

Now you can connect your computer and your BeagleBone Black via this wireless network. The advantage is that it only carries communications to and from your sail boat and can have a range of almost a mile using the right devices. Here is a website that provides an excellent example of how to configure and have two computers talk over this type of dedicated wireless link: `http://www.examples.digi.com/get-started/basic-xbee-802-15-4-chat/`.

Objective complete – mini debriefing

Now you can sail your boat, controlling it all through an external keyboard or through a ZigBee wireless network from your computer. If you want to fully automate your system, you could add your GPS and then have your sailboat sail to different positions. One additional item you might want to add for a fully automated system is a wind sensor. The following image shows a wind sensor that is fairly inexpensive from `http://www.moderndevices.com`:

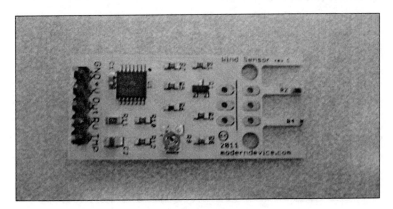

You can mount it to the mast if you'd like; I used a small piece of heavy-duty tape and mounted it to the top of the mast, similar to what is shown in the following image:

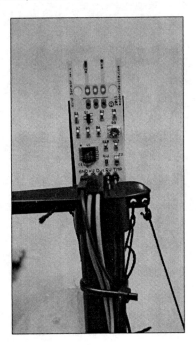

To add this to your system, you'll also need a way to take the analog input from the sensor and send it to the BeagleBone Black. You could try sending it to one of the GPIO analog input pins on the BeagleBone Black. This can be a bit tricky to program, and the ADC inputs on the board can only handle up to 1.8 Volts. If you are a bit nervous about connecting directly into the GPIO, the PhidgetInterfaceKit 2/2/2 from `http://www.phidgets.com` can help. It will actually take the analog and convert it to a reading that you can access via USB. The following image is how this device appears:

The following image shows the wind sensor connected to the converter:

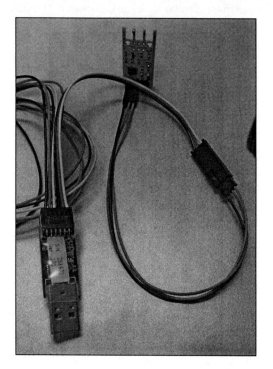

A wiring diagram is as follows:

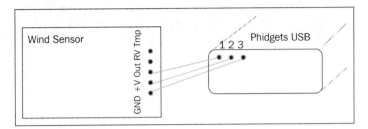

Now you can access the wind speed from the USB connection the same way as you received data from the other USB devices that you have already used. The Phidgets website will lead you through the download process; I chose Python as my language, and downloaded the appropriate libraries and example code. When I run the program I get the following output when blowing on the sensor:

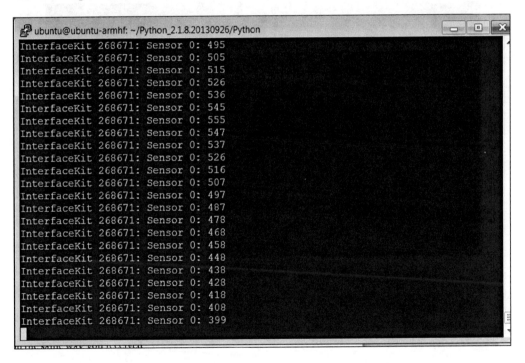

Now that you have a way to measure your location, and a way to measure the wind, you can use your BeagleBone Black to sail all by itself. You'll need to be careful with waterproofing, especially when sailing in a heavy wind. Think about attaching the hatch that covers the electronics securely; I added small screws and tabs to hold the hatch and also some waterproof sealant.

Using the BeagleBone Black in flying robots

You've now built robots that can move around on a wheeled structure, and robots that have legs, and robots that can sail. You can also build robots that can fly, relying on the BeagleBone Black to control their flight. There are several possible ways to incorporate the BeagleBone Black into a flying robotic project, but the most straightforward way is to add it to a **quadcopter** project.

Quadcopters are a unique subset of flying platforms that have become very popular in the last few years. They are a flying platform that utilizes the same vertical lift concept as helicopters; however, they employ not one but four motor/propeller combinations to provide an enhanced level of stability. The following image displays such a platform:

The quadcopter has two sets of counter-rotating propellers, which simply means that two of the propellers rotate one way; the other two rotate the other way to provide thrust in the same direction. This provides a platform that is inherently stable. Controlling the thrust of all the four motors allows you to change pitch, roll, and yaw of the device. The following diagram will be helpful:

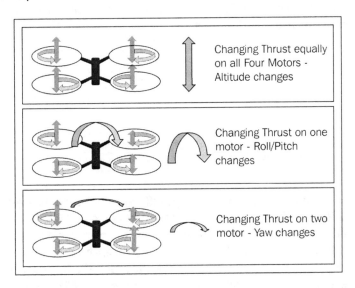

As you can see, controlling the relative speeds of the four motors allows you to control the various ways the device can change position. To move forward, or really in any direction, we would combine a change in roll/pitch with a change in thrust, so that instead of going up, the device would move forward, as shown in the following diagram:

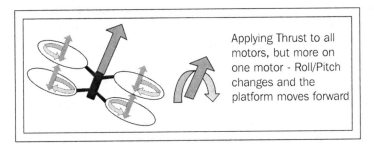

In a perfect world you might, knowing the components you used to build your quadcopter, know exactly how much control signal to apply to get a certain change in the roll, pitch, yaw, or altitude of your quadcopter. But there are simply too many aspects of your device that can vary to know this well enough to rely on a fixed set of signals. Instead, this platform uses a series of measurements of its position, pitch, roll, yaw, and altitude and then adjusts the control signals to the motors to achieve the desired result. We call this **feedback control**. The following diagram shows a feedback system:

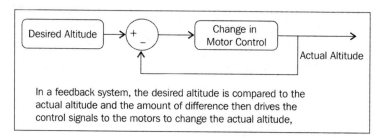

In a feedback system, the desired altitude is compared to the actual altitude and the amount of difference then drives the control signals to the motors to change the actual altitude.

As you can see, if your quadcopter is too low, the difference between the **Desired Altitude** and the **Actual Altitude** will be positive, and the **Motor Control** will increase the voltage to the motors, increasing the altitude. If the quadcopter is too high, the difference between the Desired Altitude and the Actual Altitude will be negative, and the **Motor Control** will decrease the voltage to the motors, decreasing the altitude. If the **Desired Altitude** and the **Actual Altitude** are equal, then the difference between the two will be zero, and the **Motor Control** will be held at its current value. Thus the system stabilizes even if the components aren't perfect or if a wind comes along and blows the quadcopter up or down.

One application of the BeagleBone Black in this type of robotic project is to actually coordinate the measurement and control of the quadcopter's pitch, roll, yaw, and altitude. This can be done; however, it is a very complex task, and the details of its implementation are beyond the scope of this book. There are some individuals in the open source software and hardware space working on this problem. It may well be that in the near future solutions may be available.

However, the BeagleBone Black can still be utilized in this type of robotic project by introducing another embedded processor to do the low-level control and using the BeagleBone black to manage the high-level tasks such as using the vision system of the BeagleBone Black to identify a colored ball and then guiding the platform towards it. Another option—as in the sail boat example—is to use the BeagleBone Black to coordinate GPS tracking and long-range communications via ZigBee. This is the type of example that I'll cover in this section.

Prepare for lift off

The first thing you'll need is a quadcopter. There are three approaches to this:

- ▸ Purchase an already assembled quadcopter
- ▸ Purchase a kit and construct it yourself
- ▸ Buy the parts and construct the quadcopter

In any case, to complete this section you'll need to choose one that uses the ArduPilot as its flight control system. This flight system uses a flight version of the Arduino to do the low-level feedback control we talked about earlier. The advantage to this system is that you can talk to the flight control system via USB.

There are a number of assembled quadcopters available that use this flight controller. One place to start is at `http://www.ArduPilot.com`. This will give you some information on the flight controller, and the store has several already assembled quadcopters. If you are thinking assembling a kit is the right approach, visit `http://www.unmannedtechshop.co.uk/multi-rotor.html` or `http://www.buildyourowndrone.co.uk/ArduCopter-Kits-s/33.html` as each of these not only sell assembled quadcopters but kits as well.

If you'd like to assemble your own kit, there are several good tutorials about choosing all the right parts and assembling your quadcopter. You can visit the following websites:

- ▸ `http://www.blog.tkjelectronics.dk/2012/03/quadcopters-how-to-get-started`
- ▸ `http://www.blog.oscarliang.net/build-a-quadcopter-beginners-tutorial-1/`
- ▸ `http://www.arducopter.co.uk/what-do-i-need.html`

All of the mentioned websites have excellent instructions.

You might be tempted to purchase one of the very inexpensive quadcopters that are being offered in the market. For this project you will need two key characteristics of the quadcopter:

- ▸ The quadcopter flight control will need a USB port so that you can connect the BeagleBone Black to it.
- ▸ It will need to be large enough with enough thrust to carry the extra weight of the BeagleBone Black; a battery and perhaps a webcam or other sensing devices will also be required.

No matter which path you choose, another excellent source for information is available at `http://www.code.google.com/p/arducopter`. This gives you some information on how the ArduPilot works, and also talks about Mission Planner, the open source control SW that will be used to control the ArduPilot on your quadcopter. This SW runs on the PC and communicates to the quadcopter in one of the two ways; either directly through a USB connection or through a radio connection. It is through the USB connection that you will communicate between the BeagleBone Black and the ArduPilot.

Engage Thrusters

The first step in working in this space is to build your quadcopter and get it working with an RC radio. When you allow the BeagleBone Black to control it later, you may still want to have the RC radio handy, just in case things don't go quite as planned.

When the quadcopter is flying well based on your ability to control it using the RC radio, you should then begin to use the ArduPilot in autopilot mode. To do this, download the SW from `http://www.ardupilot.com/downloads`. You can then run the SW and you should see the output that will be similar to the following screenshot:

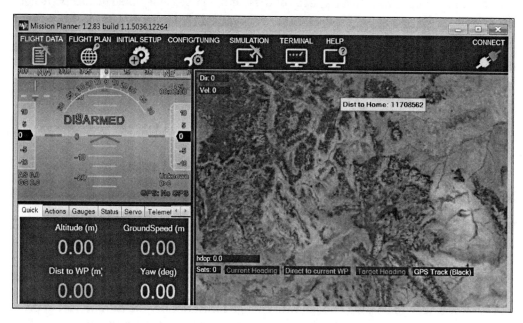

You can then connect your ArduPilot to the SW, and click on the **CONNECT** button at the upper-right corner. You should then see the output that will be similar to the following screenshot:

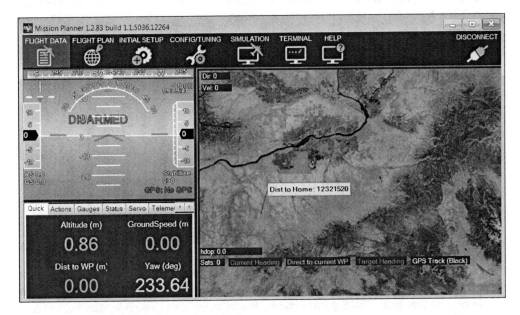

I will not walk you through how to use the SW to plan an automated flight plan; there is plenty of documentation for that on the http://www.ArduPilot.com website. Notice that in this configuration, you have not connected the GPS on the ArduPilot. What you want to do is to hook up your BeagleBone Black to the ArduPilot on your quadcopter so that it can control the flight of your quadcopter much as the Mission Planner does, but at a much lower and more specific level. You will use the USB interface, just as the Mission Planner does.

To connect the two devices, you'll need to modify the Arduino code and create some BeagleBone Black code. Then simply connect the USB interface of the BeagleBone Black to the ArduPilot and you can issue yaw, pitch, and roll commands to the Arduino to guide your quadcopter to wherever you want it to go. The Arduino will take care of keeping the quadcopter stable. Here is an excellent tutorial on how to accomplish that, albeit using the Raspberry Pi as the controller: http://www.oweng.myweb.port.ac.uk/build-your-own-quadcopter-autopilot/.

Objective complete – mini debriefing

Now that you can fly your quadcopter using the BeagleBone Black, you can use the same GPS and ZigBee capabilities mentioned in the last section to make your quadcopter semi-autonomous.

Classified intel

Your quadcopter can act in complete autonomy as well. Adding a 3G modem to the project allows you to track your quadcopter no matter where it might go, as long as it can receive a cell signal. The following image shows such a modem:

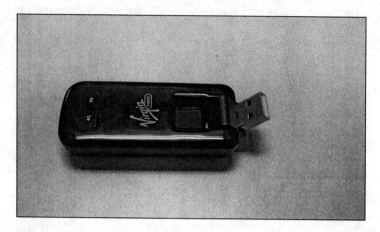

This modem can be purchased on Amazon and at your cellular service provider. Once you have purchased your modem, simply google instructions on how to configure it in Linux. An example project that puts it all together can be found at `http://www.skydrone.aero`.

Using the BeagleBone Black in submarine robots

You've explored the possibilities of walking robots, flying robots, and sailing robots. The final frontier is robots that can actually maneuver under water. It only makes sense that you can use the same techniques that you've mastered to explore the undersea world. In this section, I'll detail how to use the capabilities that you have already developed in a **Remote Operated Vehicle** (**ROV**) robot. There are, of course, some interesting challenges that come with this type of project, so get ready to get wet.

Prepare for lift off

As with the other projects in this chapter, there are possibilities to either buy an assembled robot or assemble one by yourself. If you'd like to buy an assembled ROV, visit `http://www.openrov.com`. This project, funded through Kickstarter, provides a complete package, including electronics based on the BeagleBone Black. If you are looking to build your own, there are several websites that document possible instructions for you to follow. There is one available at `http://www.dzlsevilgeniuslair.blogspot.dk/search/label/ROV`. Additionally, `http://www.mbari.org/education/rov/` and `http://www.engadget.com/2007/09/04/build-your-own-underwater-rov-for-250/` show platforms to which you can add your BeagleBone Black.

Engage thrusters

Whether you have purchased a platform or designed your own, the first step is to engage the BeagleBone Black to control the motors. Fortunately, you should have a good idea of how to do this as *Chapter 5*, *Making the Unit Mobile – Controlling Wheeled Movement*, covers how to use a set of DC motor controllers to control DC motors. In this case, you will need to control three or four motors, based on which kind of platform you build. Interestingly, the problem of control is quite similar to the quadcopter control problem. If you use four motors, the problem is almost exactly the same, except that instead of focusing on up and down, you are focusing on moving the ROV forward.

There is one significant difference: the ROV is inherently more stable. In the quadcopter your platform needed to hover in the air, a challenging control problem because the resistance of air is so small and the platform responds very quickly to changes. Because the system is so dynamic a microprocessor is needed to respond to the real-time measurements and to individually control the four motors to achieve stable flight.

This is not the case underwater where our platform does not want to move dramatically; in fact it takes a good bit of power to make the platform move through the water. You as an operator can control the motors with enough precision to get the ROV moving in the direction you want.

Another difference is that wireless communication is not available to you underwater, so you'll be tethering your device and running controls from the surface to the ROV through wires. You'll need to send control signals and video so that you can control the ROV in real time.

You have all the tools already at your disposal for this project. As already noted, from *Chapter 5*, *Making the Unit Mobile – Controlling Wheeled Movement*, you know how to hook up the DC motor controllers — you'll need one for each motor on your platform. *Chapter 4*, *Allowing the BeagleBone Black to See*, shows how to set up a webcam, so you can see what is around you. All of this can be controlled from a laptop at the surface connected via a LAN cable and running vncserver.

Objective complete – mini debriefing

Creating the basic ROV platform should open the possibility of exploring the undersea world. An ROV platform has some significant advantages. It is very difficult to lose (you have a cable attached), and because the device tends to move quite slowly, the chances for catastrophic collisions are significantly less than for many of the other projects. The biggest problem is keeping everything dry!

Mission accomplished

Now you have access to a wide array of different robotics projects that can take you over land, on the sea, or in the air. Be prepared for some challenges, and always plan on a bit of rework.

A challenge

Another possibility for an aerial project is a plane based on the ArduPilot and controlled by the BeagleBone Black. Visit `http://www.plane.ardupilot.com/` for information on controlling a fixed-wing aircraft with the Ardupilot. It would be fairly straightforward to add the BeagleBone Black to this configuration.

Index

A

Ångström 18
ArduPilot
  about 221
  URL, for documentation 221

B

basic programming constructs,
    BeagleBone Black 47-54
BeagleBone Black
  basic programming constructs 47-54
  board, accessing remotely 26-33
  board, inspecting 11
  board, plugging 12-14
  board, powering 12
  checklist 10
  connecting, to GPS device 176-188
  connecting, to mobile platform 130-137
  connecting, to USB sonar sensor 148-154
  connecting, to wireless USB keyboard 167, 168
  DC power, selecting for board 12
  display, hooking up 15-18
  files, creating 42, 43
  files, editing 42, 43
  files, saving 42, 43
  graphical user interface, adding 22-25
  keyboard, hooking up 15-18
  LEDs, blinking 13
  mobile platform, adding 103
  mouse, hooking up 15-18
  objectives 10
  operating system, modifying 18-22
  overview 9
  Python programs, running on 44-47
  USB camera, connecting to 86-89
  used, for controlling mobile platform
      programmatically 117-119
  used, for creating Python programs 44-47
  used, for flying robots 216-222
  used, for sailing robots 208-215
  using, for submarine robots 222-224
BeagleBone Black programming
  checklist 36
  features 35
  objectives 35, 36
  overview 35
board
  accessing, remotely 26-33

C

C++
  overview 54-58
cat filename command 41
clear command 41
colored objects
  detecting, OpenCV library used 97-101
commands
  interpreting, PocketSphinx used 73-80
cp filename1 filename2 command 41
Creative Labs 86

D

Dagu Rover 5 Tracked Chassis 104
DC motors 108
discus
  about 196
  installing 196

E

Emacs 42
eSpeak
  about 70
  used, for project response in robot voice 70-72

F

files
  creating 42, 43
  editing 42, 43
  saving 42, 43
filesystem
  navigating 36-41

G

general control structure
  creating, for capabilities
        communication 197-205
Global Positioning System. *See* **GPS**
GND connector 109
GPIO pins 108
GPS 176
GPS device
  accessing, programmatically 188-193
  BeagleBone Black, connecting to 176-188
GPS Receiver, for locating robot
  checklist 176
  features 175
  objectives 175
  overview 175
guvcview 86

I

ifconfig command 26
images
  viewing 86-89

K

keyboard
  used, for controlling project 169-174

L

legged platform
  checklist 126-129
  features 126
  objectives 126
  overview 125
legged robot 126
Linux commands
  cat filename 41
  clear 41
  cp filename1 filename2 41
  ll 41
  ls 41
  mkdir directoryname 41
  mv filename1 filename2 41
  rm filename 41
  sudo 41
Linux program
  creating, for controlling mobile platform 138-141
ll command 38, 41
Logitech 86
ls command 41

M

Magician Chassis 104
male-male jumper wires 107
mkdir directoryname command 41
mobile platform
  adding, to BeagleBone Black 103
  BeagleBone Black, connecting to 130-137
  checklist 104-107
  controlling programmatically, BeagleBone Black
        used 117-119
  Linux program, creating for 138-141
  motor controller, connecting to 108-116
  objectives 104
  programming, with Python 119-122
  voice command, issuing 122, 123
mobile platform speed
  controlling, motor controller used 107, 108

motor controller
about 129
connecting, to mobile platform 108-116
used, for controlling mobile platform
speed 107, 108
mv filename1 filename2 command 41

O

OpenCV
downloading 89-94
installing 89-96
OpenCV library
used, for detecting colored objects 97-101
OUTA connector 109
OUTB connector 109

P

PocketSphinx
about 73
used, for interpreting commands 73-80
Pololu
URL 106
Pololu #1372 Simple Motor Controller 18V7 106
Python
mobile platform, programming with 119-122
Python programs
creating, BeagleBone Black used 44-47
running, on BeagleBone Black 44-47

Q

quadcopter project 216
quadcopters 216

R

Remote Operated Vehicle (ROV) robot 222
rm filename command 41
Robot Operating System. *See* **ROS**
robots
flying, BeagleBone Black used 216-222
responding, to commands 80-83
sailing, BeagleBone Black used 208-215
ROS
about 206
installing 206

S

Secure Shell Hypterminal connection 27
sensors
checklist 146, 147
moving, servo used 154-158
overview 145
servo
used, for moving single sensor 154-158
servo controller
used, for connecting BeagleBone Black to mobile platform 130-137
servo motors 127
setSpeed command 122
six degrees of freedom (DOF) 127
speech functionality
checklist 62, 63
eSpeak used, for project response in robot voice
70-72
features 61
HW, hooking up 64-69
objectives 62
overview 61
PocketSphinx used, for interpreting commands
73-80
robot, responding to commands 80-83
SSH 28
submarine robots
BeagleBone Black, using for 222-224
sudo command 41
system dynamics 195

T

Tightvncserver 29
tracked platform 104

U

Ubuntu 19, 36
USB camera
connecting, to BeagleBone Black 86-89
USB-ProxSonar-EZ 146
USB sonar sensor
about 146
BeagleBone Black, connecting to 148-154

V

VIN connector 109
VirtualBox 22
vision functionality
  checklist 86
  features 85
  images, viewing 86-89
  objectives 86
  overview 85
voice command
  issuing, for mobile platform 122, 123
voice commands
  issuing, for making mobile platform truly mobile
     142, 143
voice recognition program
  modifying 122, 123

W

WinSCP 31
wireless communication, with robot
  checklist 162-166
  features 161
  objectives 162
  overview 161
wireless USB keyboard
  BeagleBone Black, connecting to 167, 168

X

XBee 212

Thank you for buying
BeagleBone Robotic Projects

About Packt Publishing

Packt, pronounced 'packed', published its first book "*Mastering phpMyAdmin for Effective MySQL Management*" in April 2004 and subsequently continued to specialize in publishing highly focused books on specific technologies and solutions.

Our books and publications share the experiences of your fellow IT professionals in adapting and customizing today's systems, applications, and frameworks. Our solution based books give you the knowledge and power to customize the software and technologies you're using to get the job done. Packt books are more specific and less general than the IT books you have seen in the past. Our unique business model allows us to bring you more focused information, giving you more of what you need to know, and less of what you don't.

Packt is a modern, yet unique publishing company, which focuses on producing quality, cutting-edge books for communities of developers, administrators, and newbies alike. For more information, please visit our website: www.packtpub.com.

Writing for Packt

We welcome all inquiries from people who are interested in authoring. Book proposals should be sent to author@packtpub.com. If your book idea is still at an early stage and you would like to discuss it first before writing a formal book proposal, contact us; one of our commissioning editors will get in touch with you.

We're not just looking for published authors; if you have strong technical skills but no writing experience, our experienced editors can help you develop a writing career, or simply get some additional reward for your expertise.

Building a Home Security System with BeagleBone

ISBN: 978-1-78355-960-2   Paperback: 121 pages

Build your own high-tech alarm system at a fraction of the cost

1. Build your own state-of-the-art security system

2. Monitor your system from anywhere you can receive e-mail

3. Add control of other systems such as sprinklers and gates

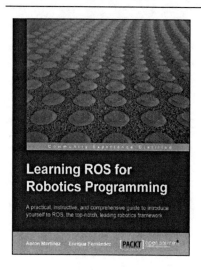

Learning ROS for Robotics Programming

ISBN: 978-1-78216-144-8   Paperback: 332 pages

A practical, instructive, and comprehensive guide to introduce yourself to ROS, the top-notch, leading robotics framework

1. Model your robot on a virtual world and learn how to simulate it

2. Carry out state-of-the-art Computer Vision tasks

3. Easy to follow, practical tutorials to program your own robots

Please check **www.PacktPub.com** for information on our titles

Robot Framework Test Automation

ISBN: 978-1-78328-303-3   Paperback: 98 pages

Create test suites and automated acceptance tests from scratch

1. Create a Robot Framework test file and a test suite

2. Identify and differentiate between different test case writing styles

3. Full of easy-to-follow steps, to get you started with Robot Framework

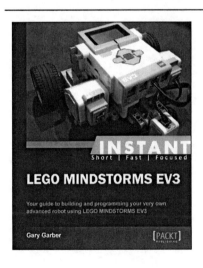

Instant LEGO MINDSTORMS EV3

ISBN: 978-1-84951-974-8   Paperback: 82 pages

Your guide to building and programming your very own advanced robot using LEGO MINDSTORMS EV3

1. Step-by-step instructions that will help you to build and program your own robot

2. Utilize all the sensors in the EV3 kit

3. Write programs with all of the essential programming commands

Please check **www.PacktPub.com** for information on our titles

Lightning Source UK Ltd.
Milton Keynes UK
UKOW03f0252210915

258952UK00001B/6/P